THE GENETICS
OF THE
TETTIGIDAE

(GROUSE LOCUSTS)

BY

ROBERT K. NABOURS

(CONTRIBUTION N° 105, DEPARTMENT OF ZOOLOGY,
KANSAS STATE AGRICULTURAL COLLEGE
AND AGRICULTURAL EXPERIMENT STATION, U.S.A.)

SPRINGER-SCIENCE+BUSINESS MEDIA, B.V

ISBN 978-94-011-8681-0   ISBN 978-94-011-9487-7 (eBook)
DOI 10.1007/978-94-011-9487-7

THE GENETICS OF THE TETTIGIDAE (GROUSE LOCUSTS) [1]

by

ROBERT K. NABOURS

CONTENTS

[1] Contribution No. 105, Department of Zoology, Kansas State Agricultural College and Agricultural Experiment Station, U. S. A.

INTRODUCTION

The various genera and species of the Orthopteran sub-family *Tettigidae* may be recognized under several names in the reports of a number of entomologists. They range from the form *Bulla* of LINNÉ (1767), whose figures are unmistakable, to the recent descriptions of SCUDDER (1900), HANCOCK (1902), and other orthoptologists. Such words as *Tettix*, *Tettiginae* and *Tettigidae* apparently were derived from "*tettix*" of Greek origin, meaning grasshopper. The common name, "Grouse locust", has probably been applied because of a fanciful resemblance of some of these insects to the grouse (*Tetraoninae*).

BIOLOGY OF THE GROUSE LOCUSTS

The Distinguishing Characteristics. The following discussion concerning the characteristics which distinguish the Grouse Locusts is largely from the detailed descriptions by HANCOCK (1902) and ROBERTSON (1915). (See Fig. 1).

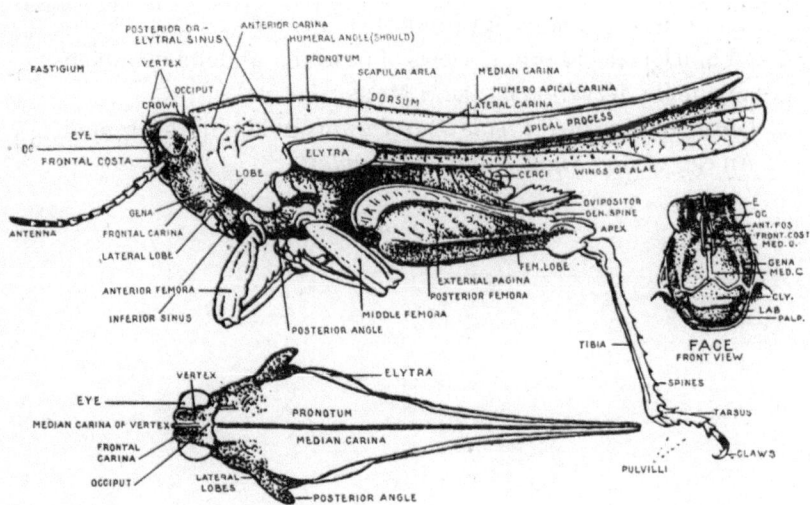

FIG. 1. Lateral, dorsal and frontal aspects of the body of a typical grouse locust, *Acrydium* (From HANCOCK)

They are among the smallest of the *Orthoptera*. The extremely developed apical process of the pronotum extends backward over the

Explanation of Plate I

(All figures from HANCOCK, 1902)

FIG. 1. *Choriphyllum foliatum* HANCOCK. Female from Jamaica.

FIG. 2. Nymph of *Acrydium obscurus* HANCOCK.

FIG. 3. *Tettigidea parvipennis* preparing the burrow for the eggs.

FIG. 4. *Acrydium granulatus* SCUDDER. Female.

FIG. 5 and 6. Dorsal and profile views of posterior abdominal appendages of a male *Acrydium hancocki* MORSE.

FIG. 7. Group of eggs of *Tettigidea parvipennis* as laid in the ground.

FIG. 8. An egg of *Acrydium ornatus triangularis*.

Plate I. — (From Hancock)

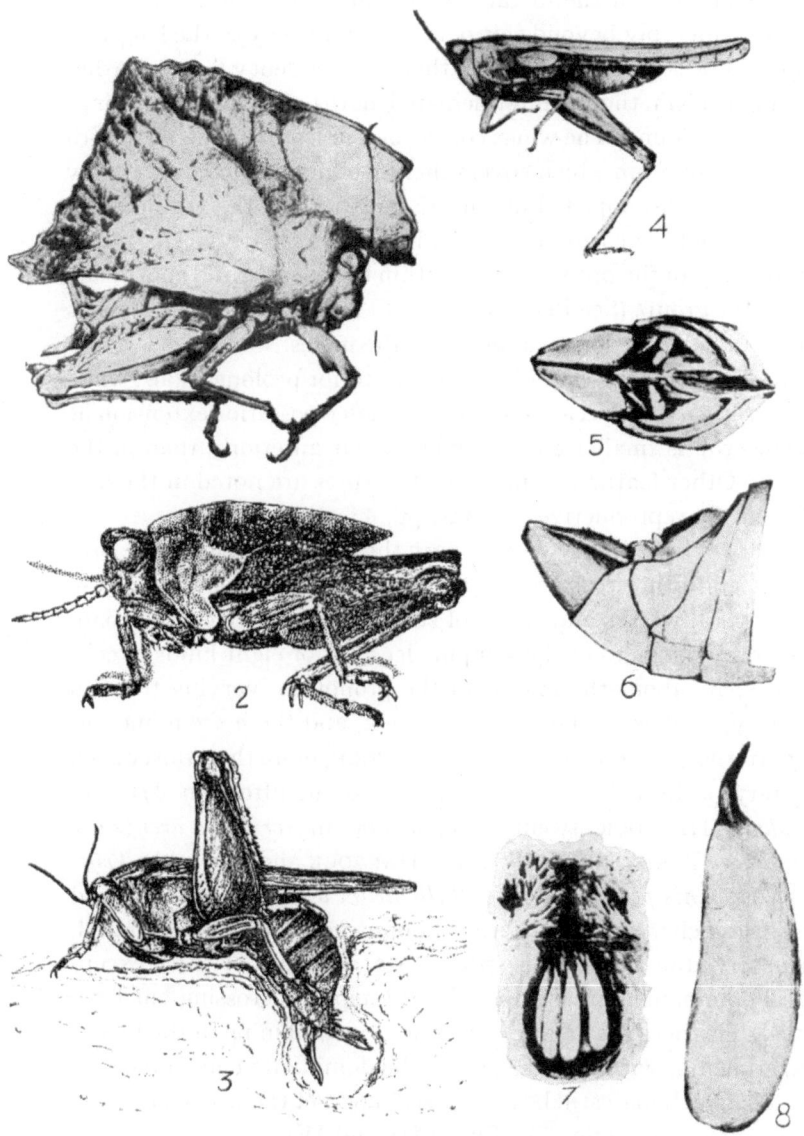

meso- and metanotum, and either back to the end of the abdomen, which coincides with the distal ends of the femora of the jumping legs, or considerably beyond this point. With respect to the length of the apical process of the pronotum, there is a tendency toward rather strict dimorphism, though intermediate lengths of pronota and wings are sometimes found. The wings correspond in length, generally, with that of the pronotum which covers them, and may be comparatively large and well-developed. Pulvilli are lacking and the tegmina are greatly reduced or absent. The individuals vary greatly in shape and size, especially of the pronota, even within the species, and enormously within the group. (See Plates I, III and IV and Fig. 1). The females are generally conspicuously larger than the males.

In the *Tettigidae* the caeca have only anterior prolongations, while in the *Acrididae* the caeca have each a smaller posterior extension as well. The crop is smaller and less rectangular anteriorly than in the *Acrididae*. Other features of internal structures are noted in the description of the reproductive systems, pp. 43—47.

There are extraordinary variations in the color patterns on the pronota, legs and other parts of the body even within a species. These include several widths and colors of stripes along the median pronotum, or on the femora of the jumping legs, six or eight kinds of conspicuous spots about the middle of the pronotum, varying through white, gray, yellow, mahogany, and black, and there are numerous other patterns. In *Paratettix texanus* HANCOCK, more than thirty such elementary patterns have been observed; about fifteen in *Apotettix eurycephalus* HANCOCK; twenty-five, or more in *Acrydium arenosum*, *A. granulatus* KIRBY, and *A. obscurus* HANCOCK and several in *Tettigidea parvipennis pennata* MORSE, *Telmatettix aztecus* SAUSSURE, and others. These elementary patterns, amounting to nearly a hundred, so far observed in the various species, are, with two or three exceptions, all dominant in inheritance. No interspecies crossing has been observed, but there are great possibilities for combining in the breeding experiments, within each species, the dominant elementary patterns so that each may still be recognized, and yet the whole be greatly different from any one. (See Plates III and IV).

It should be noted that there is a tendency for several of the same general patterns to be repeated in different genera and species. For examples, the pattern designated as C, in *P. texanus*, has a close counterpart in *Telmatettix aztecus*, and in *Nomotettix* (sp), and the K stripe of *A. eurycephalus* is almost exactly repeated in *P. texanus* and *Acrydium arenosum*. Nevertheless, these patterns are, with few exceptions, sufficiently different to render distinction fairly easy. There are also differences in the linkage relations of similar patterns in different species. For example, Y and K in *Apotettix eurycephalus* cross over to the extent of five or six per cent; while J and K, their close counterparts in *Paratettix texanus*, show no crossing over. Several other similar cases may be noted among the data.

Distribution. HANCOCK (1902) gives the distribution of the *Tettigidae* as cosmopolitan for the tropics and the temperate zones. The greater numbers and varieties of species appear to be in the tropics, and more actual species are found in North America than in Europe. HANCOCK (1902) mentioned ninety-nine species, most of them having been described in detail, and he thought there were many more. The taxonomy of the group needs a thorough revision.

The Grouse Locusts are usually found in fairly moist places in the woods, or brush, or out in open places, but most commonly on the margins of fresh water ponds and streams. Generally, as dry weather comes on, they follow the contracting margins of the ponds. As the ponds fill during wet weather, they retreat to the higher ground. Some forms as *Tettigidea parvipennis pennata*, appear to thrive best in shaded, or wooded areas, while others, as *Paratettix cuculatus* and *P. texanus* are most abundant in the open spaces. Field observations indicate that the grouse locusts would make excellent materials for ecological studies.

Food. The Grouse Locusts feed in nature mainly upon the various algae growing on the moist soils, and *Spirogyra*, *Hydrodictyon* and other filamentous algae, alive or decaying, which are left by the retreating waters of the ponds or slowly running streams. They also eat some lichens and much humus. They practically never eat grasses or other higher plants.

BREEDING HABITS

Northern species. The northern grouse locusts (Roughly from the line of the Ohio, lower Missouri, and Kansas Rivers in the U. S. A.), probably produce one, or an average of about one and one-half generations a year. The cold weather coming on in October, or November, finds both adults and various stages of nymphs and they all go into hibernation for the winter.

They make their way into tufts of grass, under stones, pieces of wood, etc., but really receive little protection from the weather. They have been observed to endure and survive -0° F. However, there is usually a high mortality, due probably as much to desiccation as to cold. There is no regularity about their going into, or emerging from hibernation. They do not become inactive till the cold weather actually arrives, and they become active during any very warm periods. During aberrantly early warm weather, they emerge from hibernation, to be driven back later in the spring if there is more cold weather.

When the warm weather arrives in the spring the adults mate and soon lay their eggs during a period of several weeks beginning in March, April or May, depending on the latitude and season. The nymphs which have hibernated over the winter contemporaneously, become adults, mate and lay eggs some weeks later. It is the supposition that the adults, hibernating over a given winter, produce a generation which in turn give a second generation which pass the following winter as nymphs. The contemporaneously hibernating nymphs, maturing and laying eggs considerably later, produce offspring which become adults in time for the coming winter — a sort of alternation of hibernating generations, the one as adults, the other as nymphs, and a generation in the summer of every other year which does not go through the winter.

Scheme of Alternation of Hibernating Generations of Northern (U. S. A.) Grouse Locusts

Winter	*Spring*	*Summer*	*Autumn*	*Winter*
↗ADULTS produce	NYMPHS become	ADULTS produce	NYMPHS remain	NYMPHS↘
↘NYMPHS become	ADULTS produce	NYMPHS become	ADULTS remain	ADULTS↗

Southern Species. In southern Texas and Louisiana, it appears that *Tettigidea parvipennis, Paratettix texanus, Apotettix eurycephalus, Telmatettix aztecus* and, perhaps other species, are active the year round except on days when the temperature is below 50° to 60° F. Thus, they are different from the northern species in that they do not spend so long a period of inactivity each year. They undoubtedly have more generations in the south than in the north, depending on the latitude. At Baton Rouge, Louisiana, males and females of *P. texanus* were entirely inactive on a frosty February morning, but were feeding and copulating at 3:00 p. m. the same day, when the temperature had risen to 70° F. There is the difference, however, between the southern Louisiana and Texas species and those of the regions of Manhattan, Kansas, and Chicago, Illinois, in that the southern individuals will breed during the winter in the greenhouse, though not so vigorously, while the northern ones will not at all till late in the winter, and then apparently better, after a period of hibernation in a cold environment.

Experimental Breeding in the Greenhouse. Some northern (region of Chicago) species of *Tettigidea, Paratettix* and *Acrydium* were first used in the experimental breeding, but as they gave only one, or at best, two generations a year, and appeared to require a critical period of hibernation, it was decided to secure some southern forms. The latter, it was thought, would find the environment of the greenhouse more like that to which they had been accustomed in nature.

Therefore, in September, 1908, some specimens of *Paratettix texanus* HANCOCK were collected near Houston, Texas, and taken to the greenhouse at the University of Chicago. They gave an average of four succeeding generations a year, and without a hibernating period. In 1910, the stocks were moved to the Kansas Agricultural Experiment Station, Manhattan, Kansas, where they have since been bred continuously, with additions of fresh specimens almost yearly, from Houston, Sugarland, San Antonio, and Austin, Texas, and Many, Louisiana. (NABOURS, 1914, 1917, 1923, NABOURS and FOSTER, 1929).

Specimens of *Apotettix eurycephalus* HANCOCK were first collected near Tampico, Mexico, in August, 1911, and others have since been added, almost yearly, from the regions of Houston, Sugarland, and San Antonio, Texas. (NABOURS, 1919, 1923, 1925).

P. texanus and *A. eurycephalus* have been bred most extensively, but *Tettigidea parvipennis pennata* and *Telmatettix aztecus* from southern Texas, and a few specimens of the latter from Pasadena, California, have been bred in considerable numbers. (BELLAMY, 1917, NABOURS and SNYDER, 1928).

Compared with *Drosophila*, the breeding of the grouse locusts is a most arduous undertaking. Two of the four succeeding generations a year are secured during the optimum season, approximately March-June, inclusive, while the other two generations are drawn out over the long period of about eight months, approximately July-February, inclusive. At times, in late summer and mid-winter, it is most difficult to keep the stocks going. The causes of the high mortality have not been ascertained, but their vitality appears to be greatly reduced during these periods.

Paratettix texanus appear to flourish for two or three generations in the greenhouse, and then, if new stocks from nature are not added, the numbers of offspring hatched decrease, sterility increases, and the mortality between the time of hatching and the time for making the first records, usually during the third instar, increases appreciably. The workers in the laboratory observe that the first and second generations of *P. texanus* from nature are much more lively and better breeders than generations more remote from nature. Some stocks have been carried continuously nineteen years, through more than sixty-five generations, but it is not thought this could have been done without the addition of individuals from nature almost yearly. It has been suggested that the decline in vigor may be due to the elimination of certain rays from the light that passes through, first, the glass roof and then during most of the day, the glass cylinders in which they are confined.

On the other hand, it is noted that the stocks of *A. eurycephalus* do not experience a similar decline in vigor and breeding activity during generations more remote from nature. In most respects, *A. eurycephalus* breed much better in the greenhouse than do *P. texanus*.

The best cages so far used consist of glass cylinders (battery jars with bottoms cut off) 8″ × 12″ and 9″ × 15″, respectively, set in bulb pots filled with sand at the bottom and a loam, rich in humus, on top. A three or four inch pot, with the hole plugged, is placed upside down in the middle and over the hole in the bulb pot before the sand and loam are put in. This inside pot is supposed to provide for better aeration of the soil. The food, consisting mainly of the filamentous algae, is placed on the projecting bottom of the inverted pot and allowed to run down the sides and over part of the soil. The cylinders have covers made of 24-mesh screen wire. (NABOURS, 1914).

A single pair, usually, is placed in the smaller cage. The eggs are laid in the ground. (Figs. 3 and 7, Plate I). Soon after hatching, the young are transferred with ordinary curved tweezers to the larger breeding cages where they remain till the records of color patterns

TABLE I. — NUMBERS TRANSFERRED COMPARED WITH NUMBERS RECORDED IN *Apotettix eurycephalus* (KANS. TECH. BULL. 17, NABOURS, 1925).

		Jan	Feb	Mar.	Apr.	May	June	July	Aug.	Sept.	Oct.	Nov.	Dec.	Yearly totals	Percent Recorded
1911	Trans.	—	—	—	—	—	—	—	—	—	180	139	20	339	10.6
	Rec.										7	28	1	36	
1912	Trans.	7	5	—	135	175	69	151	284	178	—	—	—	1004	50.3
	Rec.	6	5		40	118	69	151	72	45				506	
1913	Trans.	—	—	—	178	201	55	43	300	21	—	39	—	837	61.8
	Rec.				101	172	55	26	133	13		18		518	
1914	Trans.	52	210	617	353	584	627	232	237	410	68	87	40	3517	70.2
	Rec.	31	191	429	350	288	491	201	104	238	43	72	34	2472	
1915	Trans.	280	242	823	2252	1969	600	2124	1225	1859	3348	1008	61	15791	67.2
	Rec.	184	222	496	1527	1469	453	1460	722	1216	2438	422	13	10622	
1916	Trans.	0	288	2261	650	322	193	1453	155	411	1155	1459	22	8369	48.5
	Rec.	0	159	1742	357	142	151	539	102	70	318	459	20	4059	
1917	Trans.	180	858	2432	2282	1375	1936	3276	1583	1367	841	1451	1157	18738	60.7
	Rec.	119	401	1384	1699	918	1339	2281	1100	619	365	636	521	11382	
1918	Trans.	315	487	2498	3587	2254	3204	1528	795	99	303	3244	2078	20392	64.2
	Rec.	204	306	2282	2857	1630	1546	551	208	80	208	2084	1154	13110	
1919	Trans.	1472	1044	3945	5618	3792	5635	5574	1142	369	556	579	420	30146	73.1
	Rec.	1038	743	3135	4662	2692	4513	3548	608	189	443	266	225	22062	
1920	Trans.	531	808	2107	2540	4434	3433	2477	953	447	1593	876	926	21125	73.5
	Rec.	330	533	1783	2073	3455	2459	1968	597	251	837	598	660	15544	
1921	Trans.	898	418	2752	6175	5686	5188	3539	461	835	2549	4081	716	33298	69.9
	Rec.	666	307	2050	5106	4839	3692	2682	352	602	1214	1335	460	23305	
1922	Trans.	134	78	535	4118	5055	5177	4645	7327	2114	2625	2066	3051	36925	82.8
	Rec.	65	62	457	3550	4349	4704	4124	6702	1829	1786	1224	1740	30592	
Totals	Trans.	3869	4438	17970	27888	25847	26117	25042	14462	8110	13218	15029	8491	190481	Av. 70.4
	Rec.	2643	2929	13758	22322	20072	19472	17531	10700	5152	7659	7142	4828	134208	
	Percent Recorded.	68.3	65.9	76.5	80	77.6	74.5	70	73.9	63.5	57.9	47.5	56.8		

are made. The color patterns are distinct as early as the second instar, but records can be made better during the third instar. To wait later means the loss, in the aggregate, of a large percentage from the total of the records because the mortality is continuous and considerable.

In one experiment with *A. eurycephalus*, covering a period of twelve years, of 190,481 individuals hatched and transferred, 134,208 or 70.4 per cent survived to be recorded. Table I shows the distribution of this mortality thoughout the year and with figures 2 and 3, also indicates the months which appear to be more favorable for the breeding operations (NABOURS, 1925).

Kans. Agri. Expt. Sta. Tech. Bull. 17.

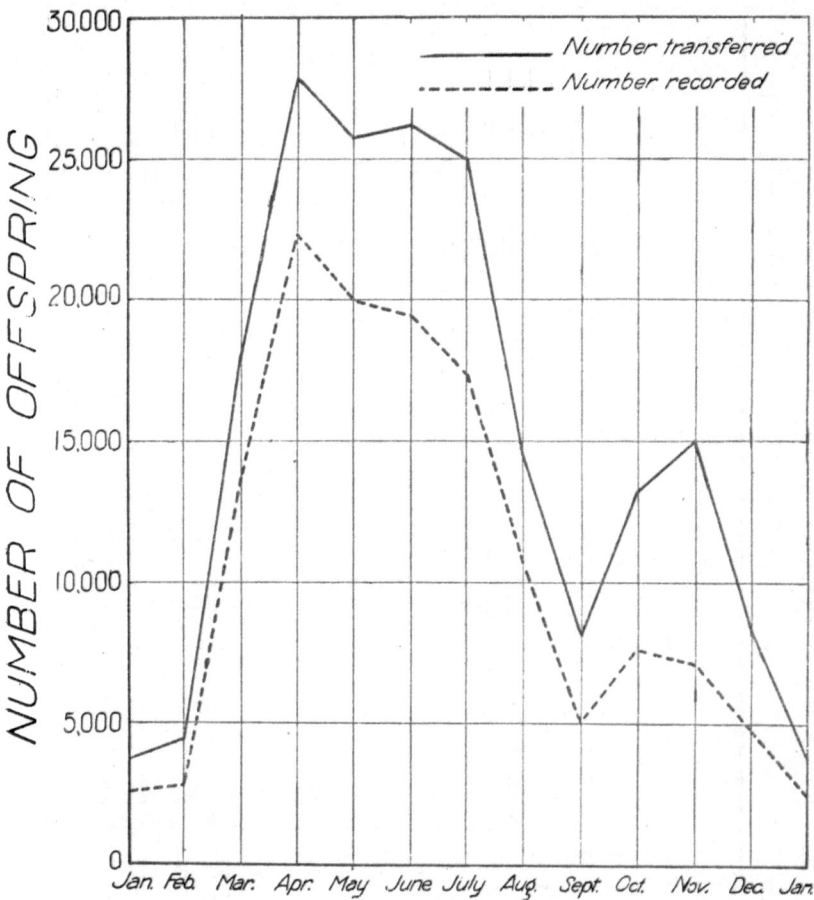

FIG. 2. Monthly totals of *Apotettix eurycephalus* offspring, 1911 to 1922 inclusive.

Kans. Agri. Expt. Sta. Tech. Bull. 17.

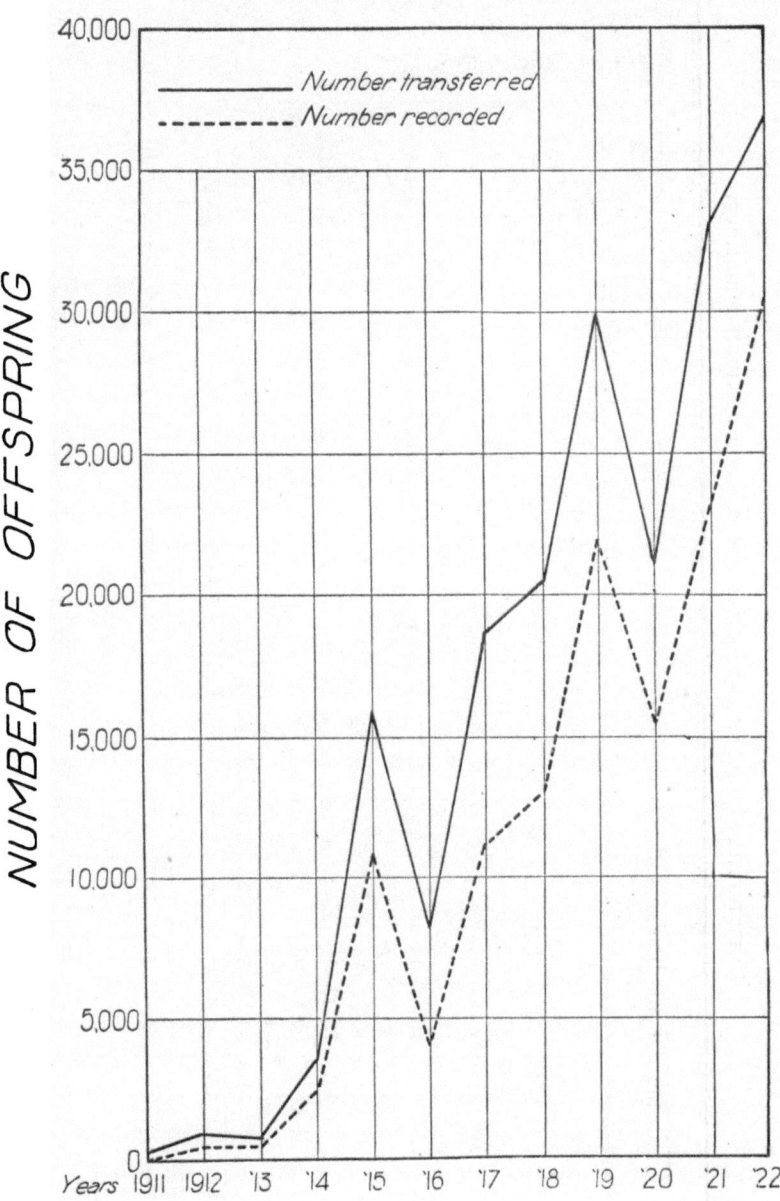

FIG. 3. Yearly totals of *Apotettix eurycephalus* offspring, 1911 to 1922 inclusive.

TABLE II. — PRODUCTIVE AND NON-PRODUCTIVE MATINGS IN *Apotettix eurycephalus* (KANS. TECH. BUL. 17, NABOURS, 1925).

(First number indicates productive, second, non-productive matings)

	Jan.	Feb.	Mar.	Apr.	May	June	July	Aug.	Sept.	Oct.	Nov.	Dec.	Yearly totals	Combined totals	Percent of matings productive
1911	—	—	—	—	—	—	—	—	10- 0	—	—	—	10- 0	10	100
1912	—	2- 0	3- 0	10- 0	1- 0	—	9- 8	0- 5	—	0- 1	—	2- 0	27- 14	41	65.8
1913	1- 2	3- 2	2- 0	—	—	5- 1	1- 0	0- 3	—	3- 1	3- 1	2- 1	17- 10	27	62.9
1914	12- 1	2- 1	9- 0	1- 0	6- 0	3- 2	9- 2	6- 2	2- 5	3- 2	6- 0	—	59- 15	74	79.7
1915	9- 2	1- 0	22- 3	7- 1	5- 0	25- 7	14- 2	16- 2	17- 3	12- 2	8- 9	—	136- 31	167	81.4
1916	—	15- 3	14- 4	8- 2	2- 2	17- 19	15- 16	4- 3	20- 8	3- 0	15- 8	—	116- 67	183	63.3
1917	18- 3	28- 3	25- 6	16- 1	27- 7	21- 4	27- 6	7- 1	6- 4	26- 5	5- 0	3- 2	223- 43	266	83.8
1918	8- 0	15- 6	24- 1	31- 3	24- 6	22- 6	8- 10	7- 25	36- 12	37- 4	28- 5	17- 3	249- 86	335	74.3
1919	38- 19	73- 13	35- 8	33- 1	56- 6	55- 14	18- 34	18- 35	12- 39	16- 15	4- 15	9- 8	372- 220	592	62.8
1920	47- 27	35- 12	52- 5	30- 2	61- 15	17- 9	23- 18	40- 24	18- 9	8- 2	22- 15	18- 10	371- 148	519	71.4
1921	23- 9	74- 16	65- 9	49- 16	50- 7	62- 65	6- 13	37- 19	85- 49	5- 16	2- 4	—	458- 223	681	67.2
1922	2- 1	37- 2	54- 8	22- 2	23- 3	48- 8	52- 13	17- 2	24- 7	54- 14	18- 2	2- 1	353- 63	416	84.8
Totals . .	158- 64	285- 58	305- 44	207- 28	255- 46	275- 135	182- 122	152- 121	230- 136	164- 61	111- 59	67- 46	2391- 920		—
Combined Totals. .	222	343	349	235	301	410	304	273	366	225	170	113		3311	—
Percent of matings productive.	71.1	83	87.3	88	84.7	67	59.8	55.6	62.8	72.8	65.2	59.2			Av. 72.2

Over the same period of twelve years, and the same experiment,
among the 3,311 matings, exclusive of parthenogenesis, in *A. eurycephalus*, 920 produced no offspring while 2,391 matings gave from 1 to
288 individuals each. Table II, and Figs. 4 and 5 indicate the distribution of the productive and non-productive matings throughout the
months of the years.

Kans. Agri. Expt. Sta. Tech. Bull. 17.

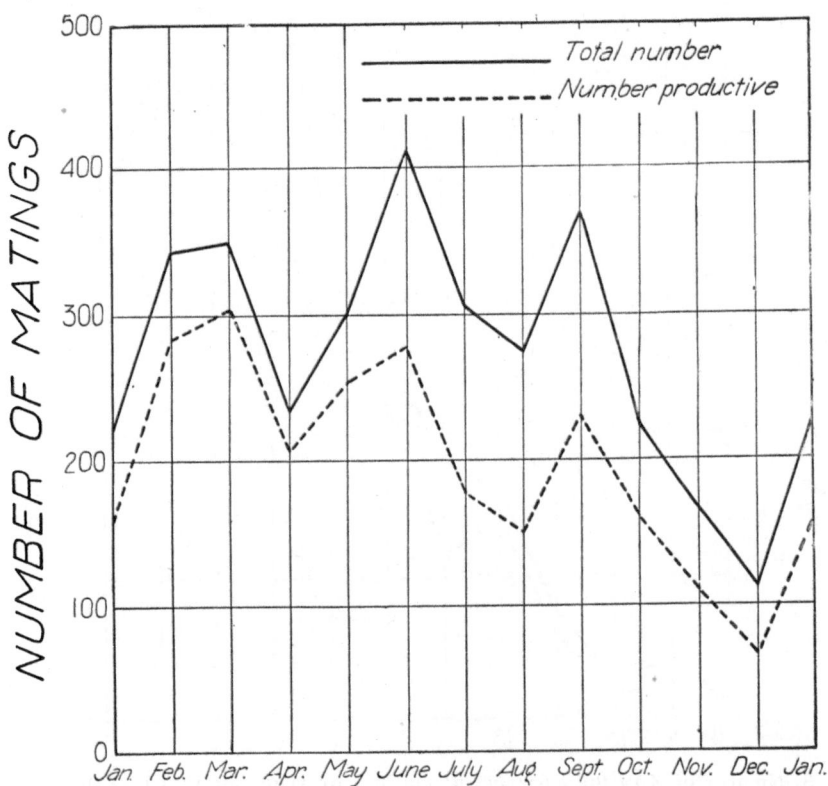

FIG. 4. Number of *Apottetix eurycephalus* matings by months, 1911 to 1922
inclusive.

Kans. Agri. Expt. Sta. Tech. Bull. 17.

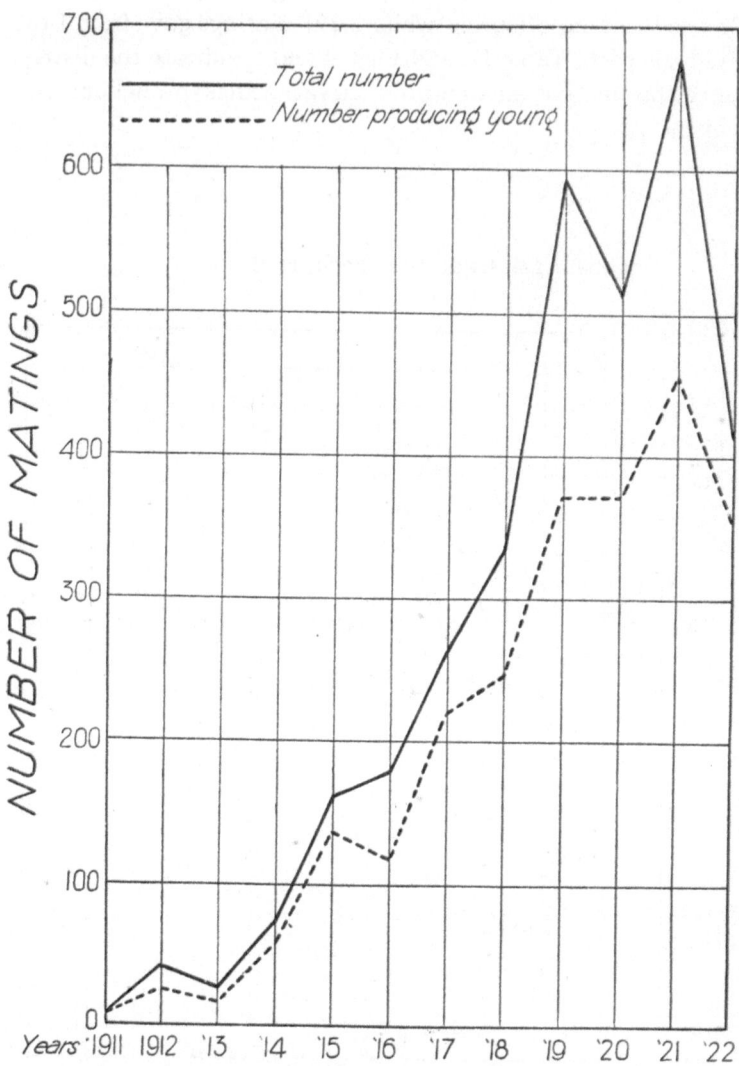

FIG. 5. Number of *Apotettix eurycephalus* matings by years, 1911 to 1922, inclusive.

Individuals often live in the laboratory as long as four to six months. The oldest recorded life-span is that of a female which lived a few days more than nine months. The hibernating individuals of the north probably frequently live eight to ten months, and longer, in nature.

After the records are made, all, except those reserved for further mating, are preserved in 70—80 per cent alcohol. After changing once into fresh 90—95 per cent alcohol, the vials are hermetically sealed with a torch. The mated individuals, after the eggs are deposited, are similarly preserved when they do not die prematurely and become stained. In this way a large number of parents and offspring have been preserved over a period of eighteen years. The patterns of those which have been in alcohol, even as long as seventeen years, except in cases where some were preserved too soon after moulting, are practically as distinct as on the day the individuals were killed.

THE REPRODUCTIVE SYSTEMS OF APOTETTIX EURYCEPHALUS

The following abbreviated account of the reproductive systems of *Apotettix eurycephalus* is almost entirely from the detailed description by HARMAN (1925).

The Male Reproductive System. Nine conical follicles, any one of which is actually independent, compose each of the pair of testes which lie close together in the mid-dorsal part of the abdomen (Fig. 3 Plate II). A follicle is composed of cysts which give it a segmented appearance. All the cells in any cyst are in approximately the same stage of development. HARMAN recognized four zones in a follicle: (1) the germarium, or zone of multiplication, (2) the zone of growth, (3) the zone of the maturation divisions, and (4) the zone of transformation. (Fig. 5, Plate II). The germarium, at the tip of the follicle, includes the primordial germ cells and spermatogonia. The successive stages of growth, with minor exceptions, are the same as those of *Paratettix texanus* described by HARMAN (1920). The zone of maturation includes the primary and secondary spermatocytes. The zone of transformation contains the successive stages in the development of the spermatids and their transformation into spermatozoa which appear to be the same as in *P. texanus* (HARMAN 1915). The functions of a seminal vesicle are apparently performed by the basal cyst of the follicle.

The paired vasa deferentia lie very close together, and each upon the ventral surface of one-half of the testicular mass. They extend posteriorly, one on either side of the alimentary canal, to enter the anterior end of the ejaculatory duct. The vasa efferentia from the follicles enter the anterior part of the respective vas deferens. (Fig. 3, Plate II).

The two masses of accessory glands are composed of about nine convoluted tubes each, and they lie ventral and lateral to the testes. (Fig. 1, Plate II). One tube in each mass is shortened and enlarged at the distal end. However, it shows no histological peculiarity. All the tubes open into a common duct which joins the ejaculatory duct ventral to the vasa deferentia. No spermatozoa were found in any of the tubes of the accessory glands.

The ejaculatory duct consists of an irregular bag ventral to the posterior part of the intestine and anus. The conical penis, a continuation of the ejaculatory duct, is situated at the base of the ninth sternum, or subgenital plate, and posterior to the suranal and podical plates. It is everted through a thin chitinous collar, the edges of which are thickened to form a rim on either margin. The distal rim has a deep V-shaped groove on the middorsal line. The opening in the penis for the passage of the semen is at the tip which is slightly bent (Fig. 7, Plate II). The subgenital plate (ninth sternum) has a groove extending from the apical end two-thirds back toward the base. (Fig. 5, Plate I). The spines of the cerci are coarse and scattered all over, while those of the podical and suranal plates are fine, and located only at the tip.

The Female Reproductive System. The ovaries, oviducts, vagina, spermatheca, spermathecal glands and ovipositor are considered. There are no collaterial glands. This account is also from the detailed description by HARMAN (1925).

The paired, triangular-shaped ovaries occupy the middorsal part of the abdomen, extend farther anteriorly than do the testes of the male and do not lie so close together. An ovary is composed of about eighteen separate tubes, each terminating in the egg calyx. HARMAN recognizes three regions, or zones, in each ovarian tube; (1) the terminal filament, (2) the germarium and (3) the vitellarium (Fig. 9, Plate II).

The terminal filaments are continuations of the sheaths anteriorly and, together, they form the ovarian ligament. The germarium is the zone of multiplication and corresponds to the germarium of the testes. It includes the primordial germ cells, the oögonia and the early stages of the growth period. The vitellarium, consisting of about ninety-five percent of the length of the ovarian tube, greatly increases in size at the beginning, and the cells are in a single row, except rarely, when two may lie side by side at the anterior end.

Anteriorly, in the vitellarium, the relative diameter of the cells is greater than the length, but with the enormous increase in size due to the accumulation of yolk, proceeding posteriorly, they become cuboid, then elongate, and finally very elongate ovoid. (Fig. 9, Plate II). The egg in formation in the anterior part of the vitellarium has the nucleus in the center. As the amount of yolk increases and it becomes larger and more elongate, in the posterior vitellarium, the nucleus comes to lie near the distal end of the egg which is then about 0.9 mm. long.

The fat-body invariably surrounds the ovaries and connects with all the tubes at various places. There is a correlation between the size of the eggs and the amount of fat; while the eggs are small the fat body is large; at the time of laying, the fat body is very small, becoming little more than strands of connective tissue a short time after the eggs are laid.

The eggs pass from the ovarian tubes through the egg calyces to the paired oviducts which unite to form the short vagina which, in turn, opens ventral to the anus (Fig. 2, Plate II).

The spermatheca, with its gland, lies dorsal to the oviducts and opens into the vagina middorsally. Spermatozoa are usually found in the lumen of the spermatheca, but no where else in the body of the female.

The ovipositor is composed of two pairs of processes, the dorsal and ventral gonapophyses, respectively. The opening of the vagina is formed at the base of the dorsal gonapophyses. Just below the vaginal orifice, between the gonapophyses, projects the pair of slender chitinous rods which HARMAN (1925) observed functioning as the egg-guide. (Fig. 6, Plate II).

Explanation of Plate II.

(Figures. 1—7, 9—11 from HARMAN; Figs. 12—15 from ROBERTSON [1]);
Figures 1—7, 9—12, 14, 15 refer to *Apotettix eurycephalus*; Figure
13 refers to *Paratettix texanus*).

FIG. 1. Accessory glands of male. C. convoluted tubules; CD, common duct; E, enlarged end of short tubule.

FIG. 2. Reproductive organs of female. LO, left ovary; OT, ovarian tube; EC, egg calyx; OD, oviduct; V, vagina; S, spermatheca; SD, spermathecal duct; G, spermathecal gland.

FIG. 3. Testes and vasa deferentia. F, follicle; VE, vas efferens; VD, vas deferens.

FIG. 4. Longitudinal section through vas efferens.

FIG. 5. Median section through a testicular follicle. G, germarium; GR, zone of growth; DI, zone of maturation divisions; TR, zone of transformation; 1SP, primary spermatocyte divisions; 2S, secondary spermatocyte divisions; SP, spermatids; MSP, spermatozoa.

FIG. 6. Posterior end of abdomen of female with chitin removed. OP, ovipositor; EG, egg guide; M, muscle; A, apodeme; V, vagina.

FIG. 7. Everted penis. CC, chitinous collar; AR, distal rim; PR, proximal rim; DG, dorsal groove; TP, tip of penis.

FIGS. 9, 10, 11. Sections through distal part of an ovarian tube, an egg of the middle part of vitellarium, and almost fully developed egg. TF, terminal filament; G, germarium; V, anterior part of vitellarium; N, nucleus of egg.

FIG. 12. Chromosomes of a spermatogonium of a bisexually produced individual. (See pp. 52,53).

FIG. 13. Chromosomes of an oogonium of a bisexually produced individual. (See pp. 52,53).

FIG. 14. Chromosomes of a brain cell of a parthenogenetically produced female. (See p. 53).

FIG. 15. Chromosomes of an ovarian follicle wall cell of a parthenogenetically produced female. (See p. 53).

---

[1]) Dr. ROBERTSON generously furnished the drawings of the chromosomes, figs. 12—15 (His figs. 26, 131, 1 and 15, respectively). Subsequently, he has changed the order of numbering the chromosomes, and decided that No. 3 in the upper gonomere of fig. 12 (His fig. 1) is probably the unpaired (sex) chromosome.

Plate II. — (From Harman and Robertson)

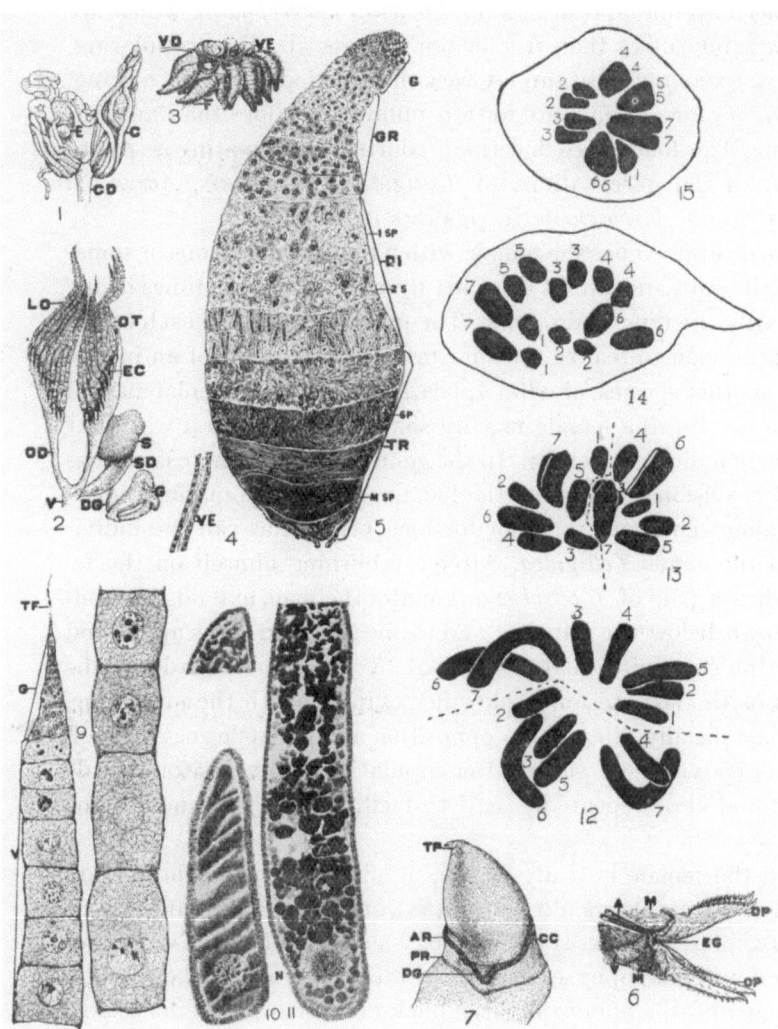

Mating, Ovipositing and Hatching. The males and females of *Tett-igidea parvipennis* spend hours, even days, at a time in copulation before and during the egg-laying period which may last several weeks. The other extreme appears to be *Apotettix eurycephalus* which are together rarely more than five or ten minutes at a time copulating. *Acrydium granulatus* also employ very short periods at a time mating. *Paratettix texanus* often mate for ten minutes to more than an hour at a time. The following concerning courtship and mating is summarized from the observations of *Tettigidea parvipennis, Acrydium granulatus* and *A. obscurus* by HANCOCK (1902).

The male approaches the female with a hurried tremulous or sometimes jerky gait, and climbs upon her from the side, sometimes facing temporarily the reverse direction. The males climb up on each other, but appear soon to realize their mistakes. At the sight of an individual of another species, or what appears to be an uncongenial male of her own species, the female usually shakes her body in a way that apparently indicates aversion. In the genus *Acrydium* there is no anatomical provision for clasping the female, so that the mating individuals cannot go about together for long periods, as can the individuals of the genus *Tettigidea*. After establishing himself on the female, when a pair of *T.parvipennis* mate, the male extends his abdomen down below the female's and to one side, and turns up the end so that the subgenital plate is affixed by its anterior border to the process of the last sternum below her ovipositor. In the meantime, the female usually offers some opposition and the male goes through a kind of convulsive orgasm. After copulation the ovipositor is made to open and close repeatedly as if to facilitate the entrance of the semen.

When the female is ready to lay, a suitable spot on the muddy ground or vegetable mould (*Paratettix*) or among moss and lichens (*Acrydium* and *Tettigidea*) is selected. The abdomen is curved under and the four gonapophyses are forced into the ground. They spread and close, and the abdomen turns back and forth on the long axis, and thus makes its way into the ground. During oviposition the front and middle pairs of legs are raised while the hind legs are drawn up to the sides, so that the abdomen extends into the ground to the maximum depth (Fig, 3, Plate I). However, there is considerable diversity in the positions assumed during egg laying; in some species, as *A. eurycephalus*, the female rests in normal position, the abdomen extending very little into the ground.

When first laid the mass of 15—40 eggs is white with a tinge of pink; they later turn to a more opaque greenish yellow-white. HANCOCK found the diameter of the eggs of *Acrydium ornatus triangularis* one-third as great as the length, which is 1.75 mm., not counting the pointed micropylar extension. The eggs of *A. eurycephalus* are about 2.1 mm. in length; those of *P. texanus* are slightly larger, while *T. parvipennis* have eggs still larger, the total length of the latter being five times the diameter (Figs. 7 and 8, Plate I). The length of the egg stage varies from 15 to 30, or more days.

The freshly hatched young are white except for the red eyes. HANCOCK describes the emergence of the young "from a little hole which the first hatched makes by worming its way to the surface." — "As if exhausted by the struggle, the young larva (nymph) on reaching the outside lies motionless for a moment; then vigorously spreading the legs and expanding the body, the veil-like amnion is torn open" — "folded backward and at last kicked off by the hind tibia". The color pattern develops rapidly and is complete within 30—40 minutes, as a rule, and is approximately the same during the rest of the life of the individual.

POLYANDRY

HANCOCK (1902) states that polyandry was frequently observed among the *Tettigidae*, mostly in the genus *Acrydium* (The *Tettix* of HANCOCK), and other grouse locusts that spent short and frequent intervals in copulation. Those that required longer periods at a time for the act of mating, such as the genus *Tettigidea*, were more likely to be restricted to the one, or fewer males for each female. Polyandry has been noted among *Apotettix eurycephalus* and *Paratettix texanus* during the fifteen and eighteen years, respectively, of their use in experimental breeding.

Polyandry in Paratettix texanus. In 1923—24, Miss CAROLINE M. PERKINS, Mr. E. H. INGERSOLL and the author carried on a series of special experiments with *P. texanus* which would indicate the possibilities of this species in polyandrous matings (NABOURS, 1927). The mere curiosity to determine how many males might take part in the parentage of the progeny of given females was the main motive. Anyway the regular program of study of gametogenesis in both the males and females would not be interrupted. If the sexual products of any male or males were to affect the color patterns or other observable features of the progeny of another male in a polyandrous mating the fact would be noted. However, the negative results of several experimenters using other materials (RABAUD 1914), and the consequences of extensive previous breeding of the grouse locusts did not render such an outcome probable.

Eight females of *P. texanus* were placed separately in mating jars, and each given opportunity to mate with several males of different dominant contrasting color patterns, and the normal recessive, of such kinds and in such ways that there would be no doubt as to the identity of the paternity of the offspring. They were exposed to males as follows: one to five different males; one to four; four, each to seven, one to nine, and one to eight males.

To five of these females the males were mated successively; that is, only one male was placed with a female at a time and, as soon as copulation occurred, he was removed and another introduced. No male was allowed to mate with a given female more than once. These five females with which four (exposed to five), three (exposed to four), seven, eight (exposed to nine), and seven males copulated successively, and only once each, respectively, gave offspring as follows: Female number 1, (+/J), 44 showing parentage of three males; female number 2, (+/Sm), 70 indicating parentage of three; female number 3, (+/C), 260 demonstrating participation by five males; female number 4, (+/+), 139 revealing five males as parents, and female 5, (+/+), 75 offspring with four males taking part (Table I, Females number 1 to number 5, NABOURS, 1927).

Three groups of seven, seven and eight males, respectively, were placed all at the same time with the other three of the separated females. They were kept under observation and notes made of the time each male spent in copulation. When the observer had to be absent, the females were segregated from the males. All seven males with female number 6 mated with her during a 14-day period; two of them once on one day each; one once each, on two days; one once each, on four days, and two on seven days each, respectively. The seventh is recorded as only trying (five times) to effect copulation and apparently not succeeding; but the observer's attention must have been diverted, at least once, for the records of the 102 offspring show that he, as well as the other six males, actually took part in the parentage. The next female, number 7, mated to six males, which were all with her over most of the period of 22 days, gave 186 offspring having the characters of four of the males. The last female, number 8, exposed to eight males, seven of them copulating with her, gave 169 offspring from four of the males. (Table I, Females number 6 to number 8, loc. cit.).

Discussion. Females numbers 3, (+/C), 4, (+/+), and 5, (+/+), gave overwhelming majorities of offspring from the males that mated with them last, +/+, on the ninth day, D/P, on the thirteenth day, and Cof/S, on the sixteenth day, respectively. Female number 7, (+/+), also gave majorities from those males, Hm/P and +/+, that mated with her on the twentieth and twenty-first days. On the other hand, female number 8, (+/+), gave more offspring from each of males +/+ and F/J, that mated with her on the twentieth and twenty-first days, respectively, than from male Cof/Sm that mated twice on the twenty-second day. The male E/I, that mated nine times on five days, and male D/P, that copulated five times on four days, did not become parents at all.

Conclusions Concerning Polyandry. 1) The last male or males copulating with a female tended to become the parents of the predominant proportion of her offspring.

2) All seven males which were placed with one female, concurrently, during a fourteen-day period, shared, respectively, in the parentage of her 102 offspring.

3) The sexual products of none of the males appeared to have any effect upon the observable characteristics of the progeny of any other males of polyandrous matings.

THE CHROMOSOMES OF THE TETTIGIDAE

The chromosome number of the *Tettigidae* was first determined in *Choriphyllum* (one species), *Nomotettix* (one species), *Acrydium* (four species), *Paratettix* (two species), and *Tettigidea* (two species). All ten species of the five genera exhibited in all cells, both somatic and germinal, from several regions of the body and gonads, the uniform number thirteen for the 2N—1 of the males and fourteen for the 2N of the females. One fundamental plan of structure of chromosomes appears to run through the entire subfamily. So regular and dependable are the size gradations in all these genera and species of the *Tettigidae* that it is possible to recognize every one of the series of six autosome pairs as well as the sex chromosomes. For instance, pairs numbers 2, 3, or 4 each would have about the same size, shape and relations in any species of any genus. (ROBERTSON 1915).

HARMAN (1915, 1920) has worked out the spermatogenesis of *Paratettix texanus* in detail, and finds 13 rod-shaped chromosomes, four larger and nine smaller ones, as the 2N—1 number. HARMAN (1925) has found in the primary spermatocytes of *Apotettix eurycephalus* six dumb-bell-shaped chromosomes and one that is ovoid. For the secondary spermatocytes, the ovoid chromosome fails to divide and there are then six and seven, respectively.

W. R. B. ROBERTSON, working in our laboratory, kindly allows the inclusion of the following observations he has lately made on *Paratettix texanus* and *Apotettix eurycephalus*. In both species there are thirteen chromosomes in the spermatogonia and various soma cells, six bivalents and the X-chromosome in the primary spermatocytes, and seven and six univalents in the secondary spermatocytes and spermatids. The size gradations and homologies of the chromosomes of these species are as regular and dependable as those of the ten species of the *Tettigidae* previously reported by him (1915) (Figs. 12 and 13, Plate II).

So far as can be ascertained the first observation of the chromosomes of a female of *A. eurycephalus* was made in our laboratory in 1917, by A. H. HERSH who was working with Dr. MARY T. HARMAN. He noted that the diploid number in at least one cell was fourteen (NABOURS, 1919, 1925). ROBERTSON (1925) confirms the observation of HERSH, and states that the regular 2N number of chromosomes in the oögonial and somatic cells of the bisexually produced females of this species is fourteen. ROBERTSON (1929MSS.) has also discovered that the numbers of chromosomes in some of the soma cells, likewise the oögonial cells of the parthenogenetically produced females of *A*.

eurycephalus and *P. texanus* appear to range from seven to fourteen, though there are actually fourteen in all of them. When the full fourteen are manifest, the members of the homologous pairs, respectively, lie together in early cell divisions, and not far apart, each from the other, in later cell generations, in such positions as to suggest the second polar body division had been inhibited.

When only seven chromosomes appear to be present in these parthenogenetically produced individuals, *each has twice the bulk in transverse thickness (broadness of the equatorial plane) of either member of the similarly numbered pair in the cells containing seven discrete pairs.* Furthermore, in these parthenogenetically produced individuals, ROBERTSON notes that in those soma and oögonial cells *containing above seven chromosomes, there is one less of the broad chromosomes for every additional one above seven.* For example, if the *apparent* number is nine, there are five of double equatorial broadness, or bulky ones, and four smaller ones in two pairs, the homologues of which always lie together, or not far from each other, in the same manner as the similar pairs do in those cells of the parthenogenetically derived females containing the fourteen discretely paired chromosomes (Figs. 14 and 15, Plate II). ROBERTSON (1929MSS.) finds the same situation in the cells of the parthenogenetically produced males which he has examined.

ROBERTSON (1915) is convinced by the results of his work with the *Tettigidae* that parasynapsis is a fact in grasshoppers, and that there are interlockings between the pairs, but there is no such intimate fusion as results in the loss of individuality or continuity as a thread. The same author finds among the chromosomes of *Syrbula* and *Chorthippus* (*Acrididae*) bodies similar to those observed by JANSSENS (1909) in *Triton* and *Batrachoseps* and upon which the chiasmatype theory was based, but there is no agreement with JANSSENS' interpretations. It is concluded that the *Tettigidae* show that the spermatogonial chromosomes on entering synapsis are already split. ROBERTSON finds in the maneuvers of the chromosomes of the *Tettigidae* and *Acrididae* (*Chorthippus* and *Jamaicana*) bases, other than the chiasmatype theory, for coupling and repulsion, the crossing over of factors observed in the breeding of animals and plants.

THE CHROMOSOMES OF OTHER ORTHOPTERA

McClung (1914) had examined over forty genera of the *Acrididae*, and with rare exceptions, the males possessed twenty-three chromosomes. It was evident that the organization of the chromosome complex was very definite and precise "in number, size and form, fiber attachment, arrangement in the metaphase and behavior during division of its elements".

In the *Locustidae*, McClung (1914) had observed in five genera, and others not identified, 33 chromosomes as the diploid number of the males, with an unusually large accessory chromosome. When the work of all the investigators of the family is considered, one is uncertain as to the uniformity of numbers of chromosomes, but in other respects, there appear to be no essential differences between the *Locustidae* and the *Acrididae*.

The *Gryllidae, Phasmidae, Blattidae* and *Forficulidae*, the latter no longer considered as belonging to the Orthoptera, have not had their chromosomes studied with so much care and detail, and there is no such unanimity of agreement on numbers, size and relations as in respect to the other families and sub-families. See summaries by McClung (1914).

General Consideration of the Chromosomes of Orthoptera. The following is evidence indicating the persistence of the chromosomes as morphological entities (McClung, 1914): (1) In each cell division the chromosomes perpetuate themselves with accuracy; (2) "not only does the individual animal show in each of its cells the same series of elements, but all the members of the species, genus and even family, with few exceptions, have it"; (3) the attachment and other relations of the spindle fibers are constant; (4) even the occasional peculiarities are invariable and uniform, as in *Hesperotettix* where there is the ever recurring union between the accessory chromosome and one of the tetrads.

McClung (1902), on discovering the accessory chromosome, suggested it might be a sex determinant.

Sutton (1902) first indicated that pairs of chromosomes may be heteromorphic with respect to other pairs and that during the maturation processes they segregate in such manner as would be required in following the Mendelian Laws of Heredity, provided the chromosomes carry the factors for the somatic characteristics.

CAROTHERS (1913, 1917, 1921), VOINOV (1914), WENRICH (1914, 1916) and ROBERTSON (1915, 1916) have noted the heteromorphic pairs of chromosomes in *Acrididae, Gryllidae* and *Tettigidae.*

CAROTHERS (1917, 1921) had determined, through the examination of a large number of wild individuals of *Hesperotettix*, and by father and sons in breeding experiments with *Circotettix*, the remarkable fact that *the members of the heteromorphic, telomitic and atelomitic pairs segregate at random in such a way as to parallel the behavior of unit characters in Mendels' first and second laws.*

INHERITANCE EXPERIMENTS

Characteristics Studied. The *Tettigidae* have long been noted for the extraordinary range and variations of their color patterns. It should be stated at the outset that neither excessive temperatures, aridity, humidity, acidity, salinity, sunlight, through glass and direct, darkness, numerous colors of soil, pots, food and excreta, starvation, fungus diseases, parasitism, nor any other observable features of environment, applied intentionally or accidentally, has ever changed any color patterns to an appreciable extent, or in any way influenced those of the progeny. Some of the elementary patterns, as B and C, in *Paratettix texanus*, have been carried in the laboratory twenty years, through more than sixty-five generations, and numerous combinations with other patterns, and yet remain exactly the same, so far as can be observed, as they were, and still are in nature. The studies of the inheritance of the variations in morphological features have not yet proceeded far enough to justify discussion of them.

The studies have thus far pertained almost exclusively to some of the dominant color patterns, and the normal recessives, of the following species: *Paratettix texanus* (NABOURS, 1914, 1917, 1923, NABOURS and FOSTER, 1929.), *Apotettix eurycephalus* (NABOURS, 1919, 1923, 1925), *Tettigidea parvipennis pennata* (BELLAMY, 1917) and *Telmatettix aztecus* (NABOURS and SNYDER, 1928).

Paratettix texanus HANCOCK — *The Color Patterns of P. texanus.*

COLOR PATTERNS OF PARATETTIX TEXANUS HANCOCK

Explanation of Plate III

The drawings have been made by Mr. S. FRED. PRINCE, and as printed, are about two and one-half times natural size. They are mostly of females which are usually larger than the males, but have the same color patterns.

All have long wings and pronota, except Cext/Cext which has intermediate length of wings and pronotum, and Cofθ/S which has short wings and pronotum.

The two top rows represent the normal recessive and twenty-three dominant elementary color patterns. The recessive pattern, sf (specked femora), is shown in the hybrid complex Esf/Jsf. The recessive color, Φ, is shown in BΦ/BΦ.

The third row and first three drawings of the fourth row represent a few simple and obvious hybrid combinations of elementary patterns.

The 4th to the 10th drawings, the last row, show the loosely linked pattern, θ, in various combinations. The last drawing, lower right shows an individual homozygous for the dominant B and the recessive Φ, respectively, and heterozygous for the dominant θ.

These drawings may be made to represent correctly many other combinations of the same patterns, respectively. For example, in the top row E/E represents also $+$/E; $+$/I represents I/I, and so on. In the bottom row, Bθ/H might represent B/Hθ, Bθ/Hθ, as well; $+$/Kθ equals K/Kθ, Kθ/Kθ; B/B$\theta\Phi$/Φ equals $+$/B$\theta\Phi$/Φ, Bθ/B$\theta\Phi$/Φ, $+$ θ/B$\theta\Phi$/Φ.

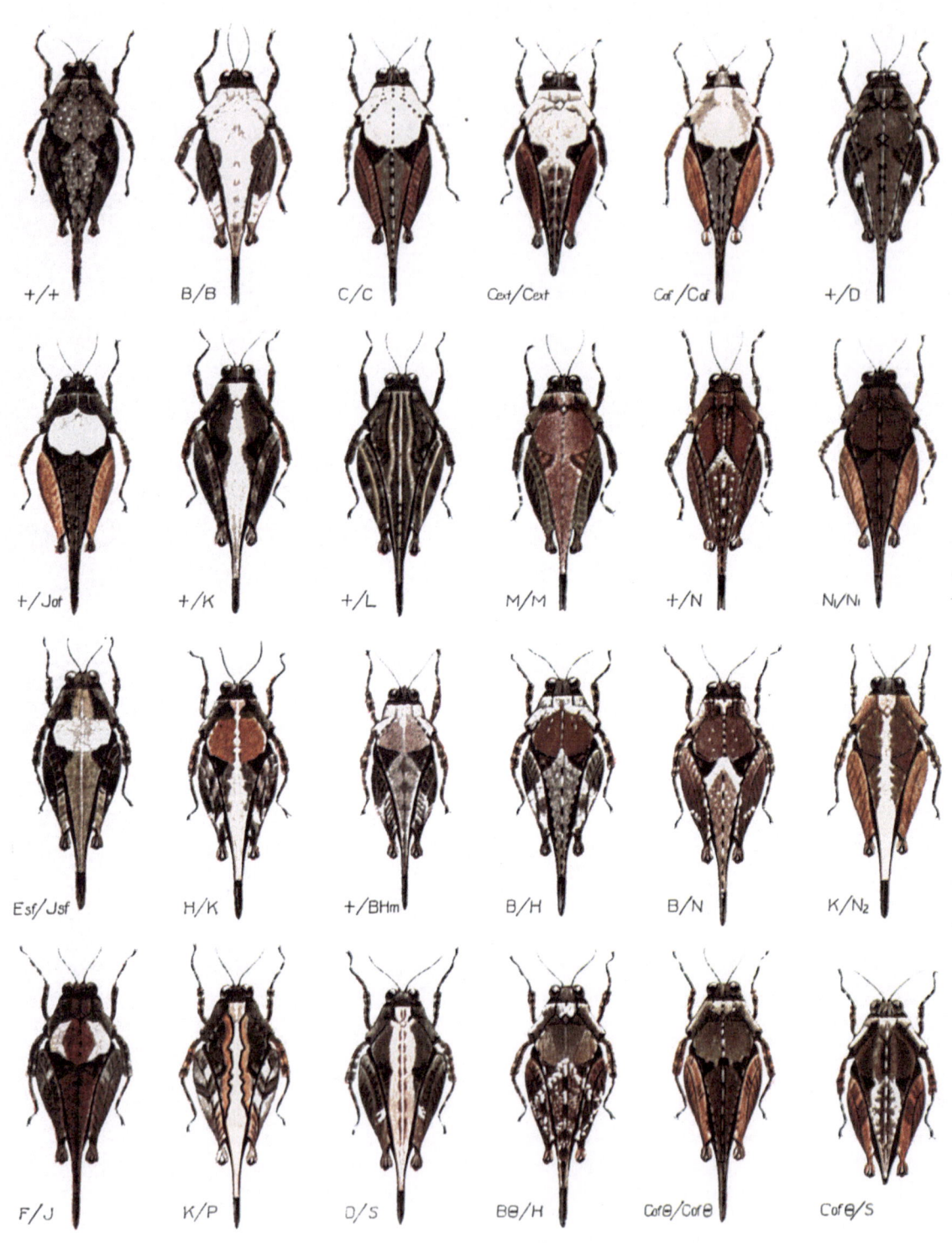

+/+ B/B C/C Cext/Cext Cof/Cof +/D

+/Jof +/K +/L M/M +/N Ni/Ni

Esf/Jsf H/K +/BHm B/H B/N K/N₂

F/J K/P D/S Bθ/H Cofθ/Cofθ Cofθ/S

COLOR PATTERNS OF

Plate III.

F/F   H/H   Hm/Hm   +/I   +/J

+/P   S/S   Sm/Sm   Sl/Sl   +/θ

C/P   B/K   J/K   K/Sm   B/I

+/Kθ   B/Kθ   J/Kθ   B/Bφ/φ   B/Bθφ/φ

ARATETTIX TEXANUS HANCOCK

Most of the elementary color patterns of *P. texanus*, and some of their hybrid complexes, have been previously approximately described and illustrated (NABOURS, 1914, 1917, 1923). The full list of those employed in experimentation up to date, some of them not included in the data, may be described, though inadequately, as follows (See Plate III): (1) +/+ (old AA), mottled gray, a pattern common to practically all the grouse locusts, and now considered the normal recessive, or "wild type"; (2) B/B, white over pronotum and parts of posterior femora; (3) C/C, white anterior pronotum, posterior dark or mottled, reddish brown legs; (4) Cext/Cext, the same as C/C, but with an extension of the white of the anterior pronotum posteriorly, and the line between the white anterior and dark posterior is not sharp; (5) Cof/Cof (old QQ) practically the same as C/C, but with red middle legs and conspicuously orange colored femora of the jumping legs; (6) D/D, the same as +/+, but with conspicuous white spots on hind femora; (7) E/E, broad yellow stripes along median pronotum and on distal ends of posterior femora; (8) F/F, broad mahogany stripes along median pronotum and posterior femora; (9) S/S, broad yellowish gray, nearly white stripes along median pronotum and on distal ends of hind femora; (10) Sm/Sm (old $\widehat{IS}\widehat{IS}$), broad brown, slightly red stripes along median pronotum and on distal ends of posterior femora (distinctly different from the other stripes and the only mutant observed to occur in the greenhouse); (11) S_1/S_1, broad nearly clear white stripes along median pronotum and on distal ends of hind femora, and with red middle legs; (12) P/P, broad brown stripes along median pronotum and on distal ends of posterior femora; (13) L/L, trilineata, three nearly white lines along the pronotum and one along femora of hind legs; (14) K/K, narrow white stripe along median pronotum, and red middle legs, almost indistinguishable from K/K, of *Apotettix eurycephalus*, (NABOURS, 1925); (15) J/J, conspicuous large white spot over broad part of pronotum, identical with Y/Y in *A. eurycephalus* (NABOURS, 1925); (16) Jof/Jof, the same as J/J, but with prominently orange colored posterior femora and red middle legs; (17) H/H, large yellow, or orange spot covering the same area as the white spot of J/J; (18) Hm/Hm, a gray, slightly orange spot, covering the same area as the spot H/H; (19) I/I, a dark mahogany spot covering the same area as that of J/J; (20) M/M, brown all over pronotum. [Hybrid M/S looks precisely like Sm/Sm. It is now thought the origin of Sm was due to the mutation of a gene closely linked with the S gene (NABOURS, 1917, pp. 48, 52, 53)]; (21) N/N, a brown gray all over; (22) N_1/N_1, dull orange, or henna all over; (23) N_2/N_2, brilliant orange all over. (These first named twenty-three factors for color patterns are extremely closely linked. Hm is the only one to have crossed over at all. Some of them may be actually multiple allelomorphs). (24) θ/θ, dense black over anterior pronotum, fading somewhat towards the posterior and extending over the hind femora; (25) sf/sf, white spots on posterior femora, resembling D/D, but recessive in heterozygotes, and not showing well, even in homozygous condition, with some of the dominant patterns, as C, Cof, Jof; (26) Φ/Φ, reddish, or pink all over, hardly discernable, almost recessive in heterozygotes. These two, sf/sf and Φ/Φ, are the only colors so far discovered in all the grouse locusts, except the normal recessive, $+/+$, that can in any sense be considered recessive, and they are only partially so. These last described three are extremely loosely linked with each other and all the others, or they may be on separate pairs of chromosomes (For θ, see HALDANE, 1920).

The breeding methods have been about the same as those employed in all the experiments with the grouse locusts (loc. cit.).

The Results of Breeding Paratettix texanus. The data are presented mainly in tables with explanations, and their use demonstrated by a few succinctly elaborated examples. In this paper, as previously (NABOURS, 1925, 1927, 1928 NABOURS and FOSTER 1929), the symbol $+$ (or $+/+$) is made to serve, as it were, a double role. When $+$ is opposite a symbol for the factor of a dominant characteristic (for example $+/B$) it refers simply to the normal recessive allelomorph of that factor. In case none of the dominant factors is present, $+/+$ represents the ensemble of indistinguishable factors that produces the mottled gray pattern, or "wild type", common to all the species of the *Tettigidae* so far studied, the pattern most frequently found in nature, and which is generally recessive to the more striking patterns.

Explanation of Table III. Twenty of the factors have shown no crossing over, any one with the other, and have thus behaved as alternatives, or multiple allelomorphs, so far as they have been submitted to breeding tests. Since the data have been assembled, another one, Cext, has been found also to be alternative to several. The other four factors, designated as Hm, θ, Φ and sf [1]) will be considered separately. The symbol + stands for the normal recessive allelomorph of the factors for dominant patterns, respectively, B, C, D, etc. Segregation in the males and females is given separately and then in totals.

The first item shows that all males heterozygous for the normal allelomorph + and the factor B gave 767 + gametes and 865 with B. All the similar females that were ever bred gave 972 + gametes and 939 B gametes. The total gametes produced by both males and females of +/B constitution was 3543, of which 1739 carried + and 1804 contained B. Following this alphabetically arranged table to B/E, it is found that of 1625 gametes produced in the breeding experiments by both males and females of this genetic composition, the males gave 376 B : 342 E, the females 453 B: 454 E, or both together 829 B: 796 E. Following on to C/I, it is found that the males gave 260 C: 288 I and the females 274 C: 247 I; both gave 534 C: 535 I, in a total of 1069 gametes. Detailed analyses of three of the matings composing part of the data of the last two examples may be seen in matings 91, 92, 93 (NABOURS, 1917). The totals at the end of the table, give the sums of the reactions of the factors, and are supposed to be unbiased because of the alphabetical arrangement. Of 110,572 gametes produced there were 27,604 : 26,881 in the males, 27,970 : 28,117 in the females; the two sexes together gave 55,574 : 54,998. This table is, in effect, a record and an exhibition of alternative gametogenesis.

---

[1]) There are not sufficient data available for this characteristic to justify a definite conclusion regarding it. However, it appears to be very loosely linked, or independent of the others.

TABLE III. SHOWING SEGREGATION OF EXTREMELY CLOSELY LINKED OR ALTERNATIVE FACTORS FOR COLOR PATTERNS IN *Paratettix texanus*

Heterozygotes	Males		Females		Males Females		Total gametes
+/B	+	B	+	B	+	B	
	767	865	972	939	1739	1804	3543
+/C	+	C	+	C	+	C	
	498	479	391	419	889	898	1787
+/Cof	+·	Cof	+	Cof	+·	Cof	
	384	424	200	212	584	636	1220
+/D	+	D	+	D	+	D	
	240	189	185	160	425	349	774
+/E	+	E	+	E	+	E	
	385	348	142	163	527	511	1038
+/F	+	F	+	F	+	F	
	135	125	242	94	377	219	596
+/H	+	H	+	H	+	H	
	42	33	248	252	290	285	575
+/Hm	+	Hm	+	Hm	+	Hm	
	9	11	77	67	86	78	164
+/I	+	I	+	I	+	I	
	232	237	131	239	363	476	839
+/J	+	J	+	J	+	J	
	756	748	292	574	1048	1322	2370
+/Jof	+	Jof	+	Jof	+	Jof	
	117	135	8	6	125	141	266
+/K	+	K	+	K	+	K	
	314	292	445	475	759	767	1526
+/L	+	L	+	L	+	L	
	53	51	79	82	132	133	265
+/M	+	M	+	M	+	M	
	65	67	64	79	129	146	275
+/N	+	N	+	N	+	N	
	122	85	216	239	338	324	662
+/N_1	+	N_1	+	N_1	+	N_1	
	85	64	92	108	177	172	349

Heterozygotes	Males		Females		Males Females		Total gametes
+/N$_2$	+	N$_2$	+	N$_2$	+	N$_2$	
	0	0	12	7	12	7	19
+/P	+	P	+	P	+	P	
	206	175	363	368	569	543	1112
+/S	+	S	+	S	+	S	
	349	312	348	344	697	656	1353
+/S$_1$	+	S$_1$	+	S$_1$	+	S$_1$	
	55	61	9	11	64	72	136
+/Sm	+	Sm	+	Sm	+	Sm	
	353	381	193	205	546	586	1132
B/C	B	C	B	C	B	C	
	2309	2221	1610	1572	3919	3793	7712
B/Cof	B	Cof	B	Cof	B	Cof	
	294	291	400	404	694	695	1389
B/D	B	D	B	D	B	D	
	22	30	39	29	61	59	120
B/E	B	E	B	E	B	E	
	376	342	453	454	829	796	1625
B/F	B	F	B	F	B	F	
	427	384	667	597	1094	981	2075
B/H	B	H	B	H	B	H	
	183	180	129	184	312	364	676
B/I	B	I	B	I	B	I	
	903	872	972	879	1875	1751	3626
B/J	B	J	B	J	B	J	
	244	253	632	643	876	896	1772
B/Jof	B	Jof	B	Jof	B	Jof	
	67	79	87	87	154	166	320
B/K	B	K	B	K	B	K	
	460	413	364	392	824	805	1629
B/L	B	L	B	L	B	L	
	175	197	126	122	301	319	620
B/M	B	M	B	M	B	M	
	187	196	45	41	232	237	469

Heterozy-gotes	Males		Females		Males Females		Total gametes
B/N	B	N	B	N	B	N	
	147	131	527	501	674	632	1306
B/N$_1$	B	N$_1$	B	N$_1$	B	N$_1$	
	12	7	13	20	25	27	52
B/N$_2$	B	N$_2$	B	N$_2$	B	N$_2$	
	131	99	78	61	209	160	369
B/P	B	P	B	P	B	P	
	428	445	327	337	755	782	1537
B/S	B	S	B	S	B	S	
	635	645	1002	973	1637	1618	3255
B/S$_1$	B	S$_1$	B	S$_1$	B	S$_1$	
	30	33	37	42	67	75	142
B/Sm	B	Sm	B	Sm	B	Sm	
	649	585	387	344	1036	929	1965
C/Cof	C	Cof	C	Cof	C	Cof	
	36	36	95	84	131	120	251
C/D	C	D	C	D	C	D	
	96	95	205	225	301	320	621
C/E	C	E	C	E	C	E	
	185	162	89	94	274	256	530
C/F	C	F	C	F	C	F	
	308	287	459	459	767	746	1513
C/H	C	H	C	H	C	H	
	126	127	5	3	131	130	261
C/I	C	I	C	I	C	I	
	260	288	274	247	534	535	1069
C/J	C	J	C	J	C	J	
	206	229	419	434	625	663	1288
C/Jof	C	Jof	C	Jof	C	Jof	
	30	28	21	25	51	53	104
C/K	C	K	C	K	C	K	
	212	252	74	61	286	313	599
C/L	C	L	C	L	C	L	
	72	70	47	29	119	99	218

Heterozy-gotes	Males		Females		Males Females		Total gametes
C/N	C	N	C	N	C	N	
	292	304	414	400	706	704	1410
C/N$_1$	C	N$_1$	C	N$_1$	C	N$_1$	
	0	0	2	3	2	3	5
C/N$_2$	C	N$_2$	C	N$_2$	C	N$_2$	
	0	1	121	98	121	99	220
C/P	C	P	C	P	C	P	
	155	155	408	380	563	535	1098
C/S	C	S	C	S	C	S	
	907	902	1004	975	1911	1877	3788
C/Sm	C	Sm	C	Sm	C	Sm	
	307	319	481	455	788	774	1562
Cof/D	Cof	D	Cof	D	Cof	D	
	48	59	78	74	126	133	259
Cof/E	Cof	E	Cof	E	Cof	E	
	252	218	86	74	338	292	630
Cof/F	Cof	F	Cof	F	Cof	F	
	18	12	37	30	55	42	97
Cof/H	Cof	H	Cof	H	Cof	H	
	46	45	55	92	101	137	238
Cof/I	Cof	I	Cof	I	Cof	I	
	40	32	64	59	104	91	195
Cof/J	Cof	J	Cof	J	Cof	J	
	295	292	420	444	715	736	1451
Cof/Jof	Cof	Jof	Cof	Jof	Cof	Jof	
	40	49	24	20	64	69	133
Cof/K	Cof	K	Cof	K	Cof	K	
	720	719	351	370	1071	1089	2160
Cof/L	Cof	L	Cof	L	Cof	L	
	20	18	175	202	195	220	415
Cof/M	Cof	M	Cof	M	Cof	M	
	55	40	24	19	79	59	138
Cof/N	Cof	N	Cof	N	Cof	N	
	0	0	66	71	66	71	137

Heterozy-gotes	Males		Females		Males Females		Total gametes
Cof/N₂	Cof	N₂	Cof	N₂	Cof	N₂	
	0	0	72	60	72	60	132
Cof/P	Cof	P	Cof	P	Cof	P	
	91	103	167	135	258	238	496
Cof/S	Cof	S	Cof	S	Cof	S	
	270	300	367	324	637	624	1261
Cof/Sm	Cof	Sm	Cof	Sm	Cof	Sm	
	168	135	97	110	265	245	510
D/E	D	E	D	E	D	E	
	73	73	45	61	118	134	252
D/F	D	F	D	F	D	F	
	6	8	0	0	6	8	14
D/H	D	H	D	H	D	H	
	60	56	13	13	73	69	142
D/I	D	I	D	I	D	I	
	0	0	12	13	12	13	25
D/J	D	J	D	J	D	J	
	212	186	213	202	425	388	813
D/Jof	D	Jof	D	Jof	D	Jof	
	8	10	34	33	42	43	85
D/K	D	K	D	K	D	K	
	138	141	116	106	254	247	501
D/L	D	L	D	L	D	L	
	56	65	20	17	76	82	158
D/M	D	M	D	M	D	M	
	66	41	75	45	141	86	227
D/N	D	N	D	N	D	N	
	10	18	63	62	73	80	153
D/N₁	D	N₁	D	N₁	D	N₁	
	0	0	1	7	1	7	8
D/P	D	P	D	P	D	P	
	136	162	183	167	319	329	648
D/S	D	S	D	S	D	S	
	138	123	214	230	352	353	705

Heterozygotes	Males		Females		Males Females		Total gametes
D/S$_1$	D	S$_1$	D	S$_1$	D	S$_1$	
	8	20	0	0	8	20	28
D/Sm	D	Sm	D	Sm	D	Sm	
	266	258	644	628	910	886	1796
E/F	E	F	E	F	E	F	
	6	19	42	50	48	69	117
E/H	E	H	E	H	E	H	
	51	59	86	83	137	142	279
E/I	E	I	E	I	E	I	
	197	207	147	125	344	332	676
E/J	E	J	E	J	E	J	
	204	198	184	182	388	380	768
E/K	E	K	E	K	E	K	
	285	279	218	287	503	566	1069
E/L	E	L	E	L	E	L	
	116	105	130	126	246	231	477
E/M	E	M	E	M	E	M	
	50	60	33	37	83	97	180
E/N	E	N	E	N	E	N	
	7	9	76	68	83	77	160
E/N$_2$	E	N$_2$	E	N$_2$	E	N$_2$	
	26	26	51	33	77	59	136
E/S	E	S	E	S	E	S	
	71	57	93	85	164	142	306
E/Sm	E	Sm	E	Sm	E	Sm	
	68	84	47	58	115	142	257
F/H	F	H	F	H	F	H	
	14	24	36	45	50	69	119
F/J	F	J	F	J	F	J	
	81	71	210	202	291	273	564
F/Jof	F	Jof	F	Jof	F	Jof	
	0	0	27	20	27	20	47

Heterozy-gotes	Males		Females		Males Females		Total gametes
F/K	F	K	F	K	F	K	
	28	19	68	73	96	92	188
F/P	F	P	F	P	F	P	
	46	49	0	0	46	49	95
F/S	F	S	F	S	F	S	
	35	43	15	7	50	50	100
F/Sm	F	Sm	F	Sm	F	Sm	
	0	0	41	49	41	49	90
H/I	H	I	H	I	H	I	
	30	37	73	82	103	119	222
H/J	H	J	H	J	H	J	
	56	57	0	2	56	59	115
H/K	H	K	H	K	H	K	
	315	262	87	77	402	339	741
H/L	H	L	H	L	H	L	
	31	37	22	39	53	76	129
H/N	H	N	H	N	H	N	
	0	0	19	23	19	23	42
H/N_2	H	N_2	H	N_2	H	N_2	
	47	41	0	0	47	41	88
H/P	H	P	H	P	H	P	
	67	68	24	26	91	94	185
H/S	H	S	H	S	H	S	
	244	215	168	192	412	407	819
H/S_1	H	S_1	H	S_1	H	S_1	
	51	57	29	57	80	114	194
H/Sm	H	Sm	H	Sm	H	Sm	
	25	30	83	70	108	100	208
I/J	I	J	I	J	I	J	
	72	83	223	199	295	282	577
I/Jof	I	Jof	I	Jof	I	Jof	
	35	43	1	3	36	46	82

Heterozy-gotes	Males		Females		Males Females		Total gametes
I/K	I	K	I	K	I	K	
	206	211	100	86	306	297	603
I/L	I	L	I	L	I	L	
	133	100	155	149	288	249	· 537
I/N	I	N	I	N	I	N	
	52	31	0	0	52	31	83
I/N_2	I	N_2	I	N_2	I	N_2	
	0	0	10	10	10	10	20
I/P	I	P	I	P	l	P	
	319	361	140	136	459	497	956
I/S	I	S	I	S	I	S	
	191	113	197	224	388	337	725
I/Sm	I	Sm	I	Sm	I	Sm	
	137	147	83	84	220	231	451
J/K	J	K	J	K	J	K	
	458	402	360	380	818	782	1600
J/L	J	L	J	L	J	L	
	80	67	86	81	166	148	314
J/M	J	M	J	M	J	M	
	73	61	6	8	79	69	148
J/N	J	N	J	N	J	N	
	606	626	541	555	1147	1181	2328
J/N_1	J	N_1	J	N_1	J	N_1	
	2	6	0	2	2	8	10
J/N_2	J	N_2	J	N_2	J	N_2	
	7	2	124	106	131	108	239
J/P	J	P	J	P	J	P	
	384	342	423	398	807	740	1547
J/S	J	S	J	S	J	S	
	336	276	360	400	696	676	1372
J/Sm	J	Sm	J	Sm	J	Sm	
	394	382	532	526	926	908	1834
Jof/K	Jof	K	Jof	K	Jof	K	
	338	302	73	93	411	395	806

Heterozy-gotes	Males		Females		Males Females		Total gametes
Jof/L	Jof	L	Jof	L	Jof	L	
	91	81	64	55	155	136	291
Jof/N	Jof	N	Jof	N	Jof	N	
	0	0	4	1	4	1	5
Jof/N$_2$	Jof	N$_2$	Jof	N$_2$	Jof	N$_2$	
	0	0	2	2	2	2	4
Jof/S	Jof	S	Jof	S	Jof	S	
	9	11	29	33	38	44	82
Jof/Sm	Jof	Sm	Jof	Sm	Jof	Sm	
	0	0	1	0	1	0	1
K/L	K	L	K	L	K	L	
	146	157	121	148	267	305	572
K/M	K	M	K	M	K	M	
	130	162	213	226	343	388	731
K/N	K	N	K	N	K	N	
	298	278	45	43	343	321	664
K/N$_1$	K	N$_1$	K	N$_1$	K	N$_1$	
	. 115	124	61	47	176	171	347
K/N$_2$	K	N$_2$	K	N$_2$	K	N$_2$	
	108	99	128	113	236	212	448
K/P	K	P	K	P	K	P	
	102	97	243	244	345	341	686
K/S	K	S	K	S	K	S	
	204	201	363	395	567	596	1163
K/S$_1$	K	S$_1$	K	S$_1$	K	S$_1$	
	0	0	45	42	45	42	87
K/Sm	K	Sm	K	Sm	K	Sm	
	134	144	247	287	381	431	812
L/M	L	M	L	M	L	M	
	12	14	0	0	12	14	26
L/N	L	N	L	N	L	N	
	7	3	5	1	12	4	15

Heterozygotes	Males		Females		Males Females		Total gametes
L/N$_2$	L	N$_2$	L	N$_2$	L	N$_2$	
	3	5	6	12	9	17	26
L/S	L	S	L	S	L	S	
	103	64	21	28	124	92	216
L/S$_1$	L	S$_1$	L	S$_1$	L	S$_1$	
	10	11	0	0	10	11	21
L/Sm	L	Sm	L	Sm	L	Sm	
	1	1	23	30	24	31	55
M/S	M	S	M	S	M	S	
	45	34	0	0	45	34	79
M/Sm	M	Sm	M	Sm	M	Sm	
	8	9	17	10	25	19	44
N/N$_1$	N	N$_1$	N	N$_1$	N	N$_1$	
	62	45	1	3	63	48	111
N/N$_2$	N	N$_2$	N	N$_2$	N	N$_2$	
	32	25	0	0	32	25	57
N/P	N	P	N	P	N	P	
	62	50	3	2	65	52	117
N/S	N	S	N	S	N	S	
	251	233	43	41	294	274	568
N/S$_1$	N	S$_1$	N	S$_1$	N	S$_1$	
	58	77	54	46	112	123	235
N/Sm	N	Sm	N	Sm	N	Sm	
	114	110	69	87	183	197	380
N$_1$/S	N$_1$	S	N$_1$	S	N$_1$	S	
	0	2	0	0	0	2	2
N$_1$/S$_1$	N$_1$	S$_1$	N$_1$	S$_1$	N$_1$	S$_1$	
	62	36	0	2	62	38	100
N$_1$/Sm	N$_1$	Sm	N$_1$	Sm	N$_1$	Sm	
	0	0	17	7	17	7	24
N$_2$/P	N$_2$	P	N$_2$	P	N$_2$	P	
	17	20	10	10	27	30	57

Heterozy-gotes	Males		Females		Males Females		Total gametes
N₂/S	N₂	S	N₂	S	N₂	S	
	90	115	17	18	107	133	240
N₂/Sm	N₂	Sm	N₂	Sm	N₂	Sm	
	3	3	73	84	76	87	163
P/S	P	S	P	S	P	S	
	72	75	44	44	116	119	235
S/S₁	S	S₁	S	S₁	S	S₁	
	3	4	6	11	9	15	24
S/Sm	S	Sm	S	Sm	S	Sm	
	0	0	134	142	134	142	276
	27,604	26,881	27,970	28,117	55,574	54,998	110,572

Explanation of Table IV Showing Linkage Relations of the Factor for Hm in Paratettix texanus. All the factors for dominant color patterns in *P. texanus* so far studied, except Hm and θ have failed to show crossing over, any one with another, as indicated in Table IV. Whether any would cross over if they were bred to the extent of millions, is, of course, not known, and no predictions can be made. Hm, as shown in Table IV, has certainly crossed over with B, L and S three times and once each, respectively. However, the percentage of crossing over is so low as to suggest very close linkage of this factor with the other twenty. It is interesting to note that three of these crossovers occurred in males and one in a female.

The arrangement is similar to that of Table III. All the pairings of Hm with the alternatives and their segregations are shown. It appears in the first item that when Hm and B were together, of 604 gametes produced by the males, 287 carried the factor for Hm, 316 the factor for B, and one the factor for the two, BHm linked. In another mating where BHm and L were paired in the male, of the seven gametes recorded, two carried BHm, four carried L and one HmL linked. In another group of matings where BHm and S were paired in the females, of the 273 gametes recorded, 121 carried BHm, 151 S, and one carried B alone, showing that Hm had crossed out from it.

TABLE IV. SHOWING LINKAGE RELATIONS OF THE FACTOR Hm IN
Paratettix texanus

Hetero-zygotes	Males		Crossovers Males	Females		Crossovers Females	Males Females			Total crossovers	Total gametes
Hm/B	Hm	B	BHm	Hm	B		Hm	B	BHm		
	287	316	1	216	217		503	533	1		1037
Hm/C	Hm	C		Hm	C		Hm	C			
	235	205		246	249		481	454			935
Hm/Cof	Hm	Cof		Hm	Cof		Hm	Cof			
	47	46		171	145		218	191			409
Hm/D	Hm	D		Hm	D		Hm	D			
	38	46		110	101		148	147			295
Hm/E	Hm	E		Hm	E		Hm	E			
	29	32		15	41		44	73			117
Hm/H	Hm	H		Hm	H		Hm	H			
	56	38		16	21		72	59			131
Hm/I	Hm	I		Hm	I		Hm	I			
	19	19		0	0		19	19			38
Hm/J	Hm	J		Hm	J		Hm	J			
	77	49		32	37		109	86			195
Hm/K	Hm	K		Hm	K		Hm	K			
	212	189		159	161		371	350			721
Hm/L	Hm	L		Hm	L		Hm	L			
	0	0		43	34		43	34			77
Hm/M	Hm	M		Hm	M		Hm	M			
	14	11		79	98		93	109			202
Hm/N	Hm	N		Hm	N		Hm	N			
	16	18		0	0		16	18			34
Hm/N_1	Hm	N_1		Hm	N_1		Hm	N_1			
	0	0		7	3		7	3			10
Hm/P	Hm	P		Hm	P		Hm	P			
	108	101		15	20		123	121			244
Hm/S	Hm	S		Hm	S		Hm	S			
	54	62		33	39		87	101			188

Heterozygotes	Males		Crossovers Males	Females		Crossovers Females	Males Females		Total crossovers	Total gametes
Hm/Sm	Hm	Sm		Hm	Sm		Hm	Sm		
	14	7		120	108		134	115		249
BHm/+	BHm	+		BHm	+		BHm	+		
	75	74		59	59		134	133		267
BHm/C	BHm	C		BHm	C		BHm	C		
	46	23		52	41		98	64		162
BHm/Cof	BHm	Cof		BHm	Cof		BHm	Cof		
	173	145		141	138		314	283		597
BHm/E	BHm	E		BHm	E		BHm	E		
	0	0		9	5		9	5		14
BHm/F	BHm	F		BHm	F		BHm	F		
	12	13		0	0		12	13		25
BHm/J	BHm	J		BHm	J		BHm	J		
	14	9		15	14		29	23		52
BHm/K	BHm	K		BHm	K		BHm	K		
	55	45		31	37		86	82		168
BHm/L	BHm	L	HmL	BHm	L		BHm	L	HmL	
	2	4	1	4	6		6	10	1	17
BHm/M	BHm	M		BHm	M		BHm	M		
	1	0		121	116		122	116		238
BHm/P	BHm	P		BHm	P		BHm	P		
	20	15		30	28		50	43		93
BHm/S	BHm	S		BHm	S	B	BHm	S	B	
	146	140		121	151	1	267	291	1	559
BHm/Sm	BHm	Sm		BHm	Sm		BHm	Sm		
	56	51		46	51		102	102		204
Totals	1,806	1,658	2	1,891	1,920	1	3,697	3,578	3	7,278

Explanation of Table V Showing Linkage Relations of θ *in Males and Females in Paratettix texanus.* The crossing over of θ with the factors B, C, Cof, D, etc., and their expedientially assumed common normal allelomorph, indicated by +, respectively, is shown for both males and females. The first column, to the left, gives the pairings of the factors with θ. The columns of figures, to the left, on each side, under males and females, respectively, show the numbers of pairings of factors, the middle columns the numbers of crossovers, and the column to the right the respective percentages of crossing over.

The first item, +θ, refers to the reactions of θ with the approximately common normal allelomorph of the extremely closely linked, some of them probably exact alternatives (except Hm), factors B, C, Cof, D, etc.

The linked BHm (see Table V) remained so in all cases where, in this linked condition, they were paired with θ. Therefore, as a matter of convenience, they are treated as one, though actually they occupy separate loci.

Examples of the Use of Table V. The first item shows that θ was paired with the opportunely assumed common allelomorph of the other factors, B, C, Cof, etc., in the males 1603 times and the females 1907. There were 394 crossovers, or 24.57 per cent, of crossing over in the males and 930, or 48.76 per cent, of crossovers in the females. Near the middle of the table, J and θ were paired 1263 times in the males, 1600 times in the females, and gave 345 crossovers, a percentage of 27.31, and 764 crossovers, a percentage of 47.75, respectively.

In the males there was a total of 16,361 pairings of θ with the closely linked and alternative factors B, C, D, etc., and their approximate allelomorph, with 4147 crossovers, a percentage of 25.34. In the females the corresponding figures are 21,733, 10,341 and 47.58, respectively.

TABLE V. SHOWING LINKAGE RELATIONS OF θ WITH OTHER FACTORS IN THE MALES AND FEMALES IN *Paratettix texanus*

Pairs of factors	Males			Females		
	Numbers	Cross-overs	Per cent of crossing over	Numbers	Cross-overs	Per cent of crossing over
$+\theta$	1603	394	24.57	1907	930	48.76
$B\theta$	1878	426	22.68	3866	1809	46.79
$BHm\theta$	165	36	21.81	298	145	48.65
$C\theta$	741	187	25.23	1068	512	47.94
$Cof\theta$	1559	435	27.90	1654	759	45.88
$D\theta$	202	59	29.20	373	196	52.54
$E\theta$	897	248	27.64	789	383	48.54
$F\theta$	488	147	30.12	1218	574	47.12
$H\theta$	425	100	23.52	449	221	49.22
$Hm\theta$	212	49	23.11	177	77	43.50
$I\theta$	480	102	21.25	379	169	44.59
$J\theta$	1263	345	27.31	1600	764	47.75
$Jof\theta$	355	78	21.97	379	181	47.75
$K\theta$	3120	771	24.71	3054	1471	48.16
$L\theta$	853	211	24.73	631	311	49.28
$M\theta$	158	45	28.48	467	225	48.17
$N\theta$	23	7	30.43	15	7	46.66
$N_1\theta$	60	13	21.66	72	34	47.22
$N_2\theta$	179	47	26.25	130	52	40.00
$P\theta$	384	90	23.43	420	175	41.66
$S\theta$	1143	319	27.90	2106	1023	48.57
$S_1\theta$	7	0	0.0	79	43	54.43
$Sm\theta$	166	38	22.89	602	280	46.51
Totals	16,361	4,147	25.34	21,733	10,341	47.58

Discussion of the Relations of the Factor θ. In the previous report on the inheritance of the factors for color in *Paratettix texanus* (NABOURS 1917), the factor θ was treated as independent of the others which were extremely closely linked, if not multiple allelomorphs. It might have been on a separate pair of chromosomes, though not definitely stated in such terms. However, J. B. S. HALDANE (1920) demonstrated with my data (NABOURS, 1914, 1917) that the factor θ was apparently not independent of the other factors, but was, in fact, loosely linked with them. He showed that, on an average, there was approximately 24 per cent of crossing over in the males and about 46 per cent in the females of the factor θ with the group of alternatives.

Subsequent extensive breeding of *P. texanus*, including the factor θ with the others named, has increased the data available for, and bearing on this matter more than twelve and one-half times beyond that used by HALDANE. A general average of the much greater data, shown in Table V, seems to confirm HALDANE's conclusions. The average crossing over in the males is shown to be 25.34 per cent which is 1.7 per cent greater than that indicated by the lesser data to which he had access. In the females there is shown a crossing over percentage of 47.58 which is a little more than one per cent greater than that found by HALDANE, with less than one-thirteenth the data now available.

As a result of these considerations, first suggested by HALDANE's review (loc. cit.), I have come to regard the factor θ as being loosely linked, about 25 per cent in the males and 47 per cent in the females, with the group of twenty and more closely linked, or alternative factors, and all as being on one pair of chromosomes.

Explanation of Table VI Showing the Relations of Φ in Paratettix texanus. Unfortunately the only data available are for males; the few females bred during the short time this characteristic was among the stocks were homozygous for Φ, or otherwise unavailable. The relations of Φ to θ, 48.6 per cent, in a small number is about the same as the average of its relations to the alternatives B, C, J and S, with which θ has a linkage relation, in the males, of 25.34 per cent. The first column indicates the factors that were paired, the second the numbers of pairings, the third the numbers of crossovers, or segregations of Φ, and the last column gives the percentages of segregations or crossing over.

COLOR PATTERNS OF APOTTETTIX EURYCEPHALUS HANCOCK

Explanation of Plate IV
(From Kansas Technical Bulletin 17)

The drawings have been made by Mr. S. FRED PRINCE and, as print-ed, are about two and one-half times natural size.

The top row represents the normal recessive type, $+/+$, and eleven dominant elementary patterns.

The second row, except the last one to the right, represents vari-ous combinations of the elementary dominant patterns in pairs. The last one of the second row and the first nine of the third row show three-pattern combinations. The last three of the third row and the first two and the fourth of the last row are four pattern complexes. In the fourth row, the third and fifth show five patterns, while the sixth represents a six-pattern complex.

The Y/YT, Y/T, and YT/T drawings of the fourth row show the influence of single and double doses of Y and T, respectively.

The last three of the fourth row show (1) an individual with short pronotum and wings, (2) one with the pronotum turned up and forward, and (3) an individual with the pattern obscured by an alga and intermediate length of pronotum and wings.

These drawings actually represent many other combinations of the same patterns, respectively. For example, in the top row, $+/Y = Y/Y$, $Z/Z = +/Z$ and so on. Second row: $Z/K = +/ZK$ $= Z/ZK = ZK/K = ZK/ZK$; $RK/RK = R/K = +/RK = R/RK$ $= RK/K$. In the third row: $+/MZK = M/ZK = MZ/K = MK/$ $Z = MZK/MZK = M/MZK = MZ/MZK = MZ/ZK = MK/MZK$ $= MZ/MK$. In the fourth row: $MRT/ZK = MZK/RT = MZT/RK$ $= MRK/ZT = MRT/MZK = MRK/MZT$. Double doses of none of the factors appear to make perceptible differences in the pattern complexes, except in the case of the redness of T, as shown Y/YT, Y/T, and YT/T, in the last row. These show the relative effects of the same genes in single and double quantities.

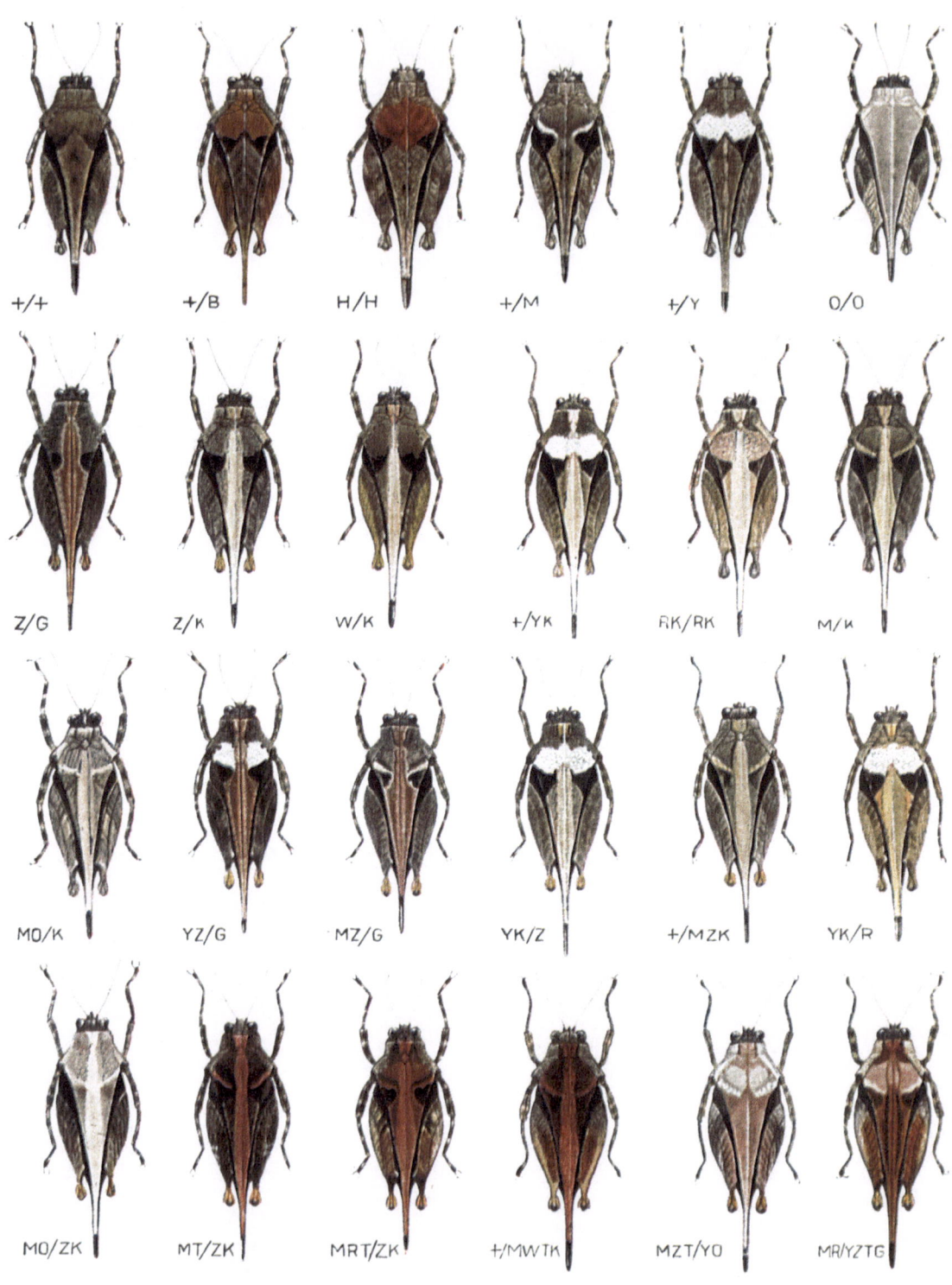

+/+   +/B   H/H   +/M   +/Y   O/O

Z/G   Z/K   W/K   +/YK   RK/RK   M/K

MO/K   YZ/G   MZ/G   YK/Z   +/MZK   YK/R

MO/ZK   MT/ZK   MRT/ZK   +/MWTK   MZT/YO   MR/YZTG

PLATE IV

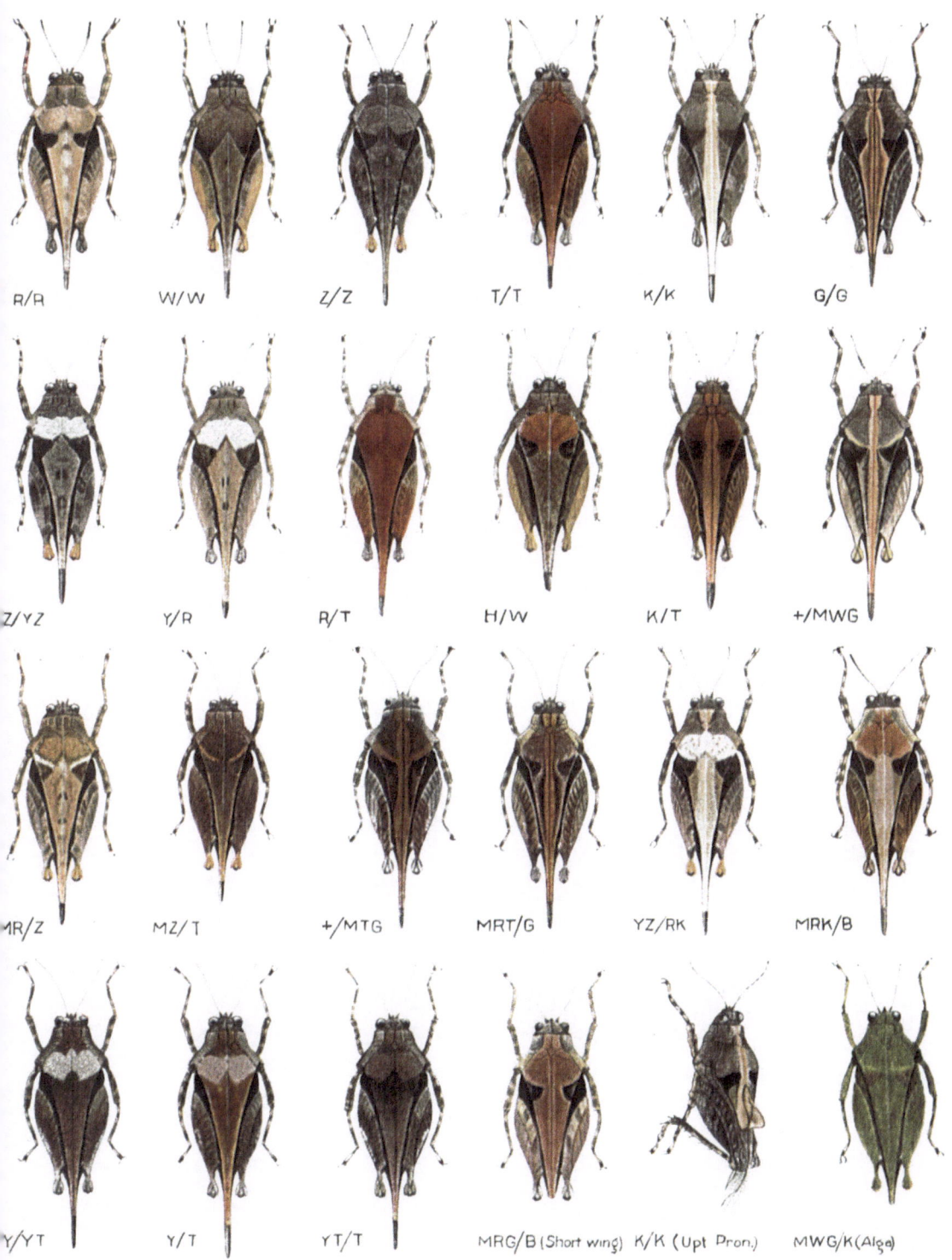

R/R W/W Z/Z T/T K/K G/G

Z/YZ Y/R R/T H/W K/T +/MWG

MR/Z MZ/T +/MTG MRT/G YZ/RK MRK/B

Y/YT Y/T YT/T MRG/B (Short wing) K/K (Upt Pron.) MWG/K (Alga)

TABLE VI. SHOWING THE RELATIONS OF Φ IN MALES OF
Paratettix texanus

Factors	Numbers	Crossovers [1]	Per cent of crossing over [1]
$\theta\Phi$	220	107	48.6
BΦ	537	252	46.92
CΦ	187	95	50.8
JΦ	91	51	56.0
SΦ	217	97	44.7
Total	1252	602	48.08

Apotettix eurycephalus HANCOCK

The Color Patterns of A. eurycephalus. (See plate IV.) (1) The symbol
+, or +/+, is used to designate the type that constitutes the majority
of the individuals found in nature. It is a mottled gray and is common
to all species of the *Tettigidae* so far studied. It is recessive to, and results from crossing over among the factors for dominant color patterns;
(2) B, brown anterior pronotum and femora of jumping legs; (3) H,
large brown spot near anterior end of pronotum; (4) M, oblique, transverse white stripes just anterior to the black triangles; (5) Y, bright
white spot over the same area as H, and almost identical with J in *P.
texanus* (Plate III); (6) Ymf, identical with Y, except that the femora
of the hind legs are brown mahogany (Ymf is not included in the table,
but has been found to be alternative with Y); (7) O, cream-all-over; (8)
R, yellowish-all-over; (9) W, yellow stripes along femora of jumping
legs, with tinge of yellow over entire body; (10) Z, prominent yellow
tips on distal ends of posterior femora; (11) θ black over anterior pronotum and extending back on posterior femora, almost exactly like θ
in *P. texanus* (So closely linked with Z, that the two, linked when
brought in from nature, have shown no crossing over. Not included
in table); (12) T, red-all-over the pronotum and jumping legs; (13)
K, narrow white, tinged with yellow, stripe along median pronotum,
slightly gray-all-over, closely resembling K, of *P. texanus* (Plate III);

---

[1] Probably independent segregation.

(14) G, reddish brown stripe along median pronotum, resembling P, of *P. texanus*, though slightly narrower (See Plate III).

Besides the prominent, mostly dominant color characteristics enumerated above, there are undoubtedly many others, most of them too subtle for ordinary observation to detect. Numerous cases of upturned, twisted and otherwise aberrant pronota, and abnormal abdomen, such as fusion of the sterna, have occurred. Some of these were apparently inheritable (NABOURS, 1925), but were left for further study.

Breeding Results in Apotettix eurycephalus (Crossing over occurs almost exclusively in the females). A large series of detailed results of the breeding of *A. eurycephalus* is included in Kan. Tech. Bull. 17, NABOURS, 1925, using all the factors for color patterns described above and illustrated in the plate (Plate IV), except the factors for Ymf and θ. There is also a table of primary data, (Table VI, NABOURS, 1925), showing the locus to locus crossing over in the females, and from which the diagram of chromosome relations, or chromosome map, was constructed. These data provided a factor map, or diagram as follows:

```
                                                      O
B                                                     R    G
H                                                     W    K
M        Y                                            Z    T
├────────────────────────────────────────────────────┼────┼┼─
0       1.3(?)                                        6.9  7.4 7.43
```

This table is summarised as Table VII in this paper.

Explanation of the Use of Table VII, Showing Locus to Locus Crossing Over Data (Primary) in Females of Apotettix eurycephalus. Mating 1906 (NABOURS, 1925), may be used for illustration. A $+/T$ male was crossed with an MRK/WG female with the following result: Non-crossovers, $+$/MRK 22, $+$/WG 23, MRK/T 19, WG/T 17; crossovers, $+$/MWG 1, $+$/MRG 1, $+$/RK 1, $+$/WK 2, RK/T 2, MRG/T 1. When the results of this mating are used as primary data, in map making, there are considered only the relations of M to R, R to W, W to G and G to K. Among the 89 individuals there were four crossovers between M and R, and four between W and G, with none between R and W, and G and K, respectively. This measures, when there are sufficient numbers, the respective distances. Since the pairs RW and GK, respectively, are alternatives, the two distances measured are those from M to RW and RW to GK.

A summary of Table VII (NABOURS, 1925), showing the total crossing over data resulting from the breeding of *A. eurycephalus* is appended below as Table VIII in this paper. These total data may not be employed in making a map, since this method differs from that of the primary data in that in measuring the distances some of the single set of data would be used twice (NABOURS, 1925, pp. 217, 220).

Explanation of the Use of Table VIII, Showing the Total Crossing Over Data (Primary plus Secundary) in Females of A. eurycephalus.

The same mating, 1906 (NABOURS, 1925), is used as follows for the total crossing over data composing the table below (Table VIII). Referring again to the mating 1906 (loc. cit.), described above, there were ten opportunities for reactions, one factor with another; M with R, K, W and G; R with K, W and G; K with W and G; and W with G. According to this arrangement the following crossovers took place in mating 1906: M crossed over from R and K and joined W and G four times; G crossed over from W and joined M and R four times; while, reciprocally, K crossed over from M and R and took position with W the same number of times. This makes four crossovers each between M and R, and M and W; eight between M and K, and M and G, four between R and K, R and G, W and K, W and G, respectively, and none between R and W, or K and G. The situation may be illustrated by a diagram as follows:

In the way described in the last paragraph above the total numbers of pairings, numbers of crossovers and percentages shown in the table were derived.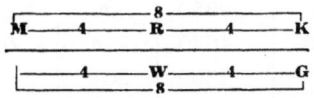

For example, item 1: Table VIII, in 323 pairings of the factors H and Y, there was no crossing over. In 14,977 pairings of M and K, there were 960 or 6.4 per cent, crossovers. In 6600 pairings of the factors for the patterns G and K, there was no crossing over.

TABLE VII. SHOWING THE LOCUS TO LOCUS CROSSING OVER DATA (PRIMARY) IN FEMALES OF *Apotettix eurycephalus*

Factors	Number	Cross-overs	Per cent	Factors	Number	Cross-overs	Per cent
HY	323	0	0	YK	836	56	6.7
HO	463	80	17.2	OR	3138	0	
HR	2142	248	11.6	OW	2876	0	
HZ	1294	64	4.9	OZ	3530	0	
HG	42	1	2.4	OT	3383	20	.6
HK	483	3	.6	OG	2786	7	.2
BM	2141	0	0	OK	1222	8	.6
BY	386	0	0	RW	9418	0	
BO	395	18	4.5	RZ	8469	0	
BW	737	64	8.7	RT	9286	32	.3
BZ	125	7	5.6	RG	3951	67	1.7
MY	3420	9	.3	RK	6611	31	.5
MO	4335	525	12.1	WZ	187	0	
MR	16299	934	5.7	WT	7364	26	.3
MW	2337	73	3.1	WG	6825	72	1.0
MZ	4094	225	5.4	WK	4980	15	.3
MT	1503	65	4.3	ZT	8913	72	.8
MG	140	20	14.3	ZG	3970	30	.8
MK	722	13	1.8	ZK	8579	33	.4
YO	8387	398	4.7	TG	8346	3	.03
YR	9850	430	4.4	TK	11652	3	.02
YW	4040	244	6.0	GK	6600	0	
YZ	5860	281	4.8				
YT	2438	213	8.7				
YG	259	10	3.9				

Summary and Discussion of the Inheritance of Color Patterns in A. eurycephalus. The three- and four-point data, and the locus to locus crossing over, or primary data (NABOURS, 1925) and Table VII, are entirely consistent in lining up the factors as represented in the diagram, p. 50, B. H, M, first; Y second; the O, R, W, Z alternatives

TABLE VIII. SHOWING THE TOTAL CROSSING OVER DATA (PRIMARY PLUS SECONDARY) IN FEMALES OF *Apotettix eurycephalus*

Factor	Number	Cross-overs	Per cent	Factors	Number	Cross-overs	Per cent
HY	323	0	0	YO	8387	398	4.7
HO	463	80	17.3	YR	11312	458	4.0
HR	2465	248	10.0	YW	8800	543	6.2
HZ	3396	223	6.6	YZ	9805	392	4.0
				YT	10480	661	6.3
HT	491	16	3.2	YG	8678	532	6.1
HG	1044	237	22.7	YK	17799	967	5.4
HK	1602	11	.7	OR	3138	0	0
BM	2141	0	0	OW	2876	0	0
BY	386	0	0	OZ	3530	0	0
BO	395	18	4.5	OT	5486	26	.5
BR	2385	62	2.6	OG	5393	14	.2
BW	1025	82	8	OK	5439	24	.4
BZ	125	7	5.6	RW	9418	0	0
BG	2828	155	5.5	RZ	8469	0	0
BK	707	51	7.2	RT	13470	60	.4
MY	3431	9	.3	RG	9988	145	1.4
MO	5013	538	10.7	RK	23712	151	.6
MR	19031	1194	6.3	WZ	187	0	0
MW	9331	366	3.9	WT	7364	26	.3
MZ	9824	653	6.6	WG	9030	81	.9
MT	16480	894	5.4	WK	10130	83	.8
MG	12344	947	7.6	ZT	8913	72	.8
MK	14977	960	6.4	ZG	5179	56	1.08
				ZK	14924	88	.6
				TG	8346	3	.03
				TK	12816	3	.02
				GK	6600	0	0

third; T fourth, and G and K fifth (Table VII). Even the summarized primary plus secondary, total data, Table VIII, are not inconsistent with this arrangement, and, in fact, if used in making a map would, make only very slight changes in linkage distances from those in the one derived from the locus to locus, or primary data. The recently discovered θ, extremely closely linked with Z (not included in the table), makes a sixth locus. These are all actually closely linked, the extremes being, on the average less than 8 per cent apart, and they are all clearly confined to the one pair of chromosomes. Also, the new factor for Ymf appears to be among the B, H, M, Y closely linked group.

The results of the breeding of *A. eurycephalus*, although only one pair of chromosomes and comparatively few factors are involved, parallels the results obtained in the breeding of *Drosophila*. The *Linear Hypothesis* (MORGAN, MULLER, BRIDGES and STURTEVANT) has proved to be the most available and explicit premise for the quantitative arrangement of the results obtained.

Telmatettix aztecus SAUSSURE. — *The Color Patterns of T. aztecus.* Specimens of grouse locusts, identified as *Telmatettix aztecus* SAUSSURE, were secured along Shoal Creek, Austin, Texas, in August, 1922. By breeding analyses it was determined that there were a recessive and four dominant elementary patterns as follows: (1) A mottled gray recessive resembling the +/+ of *Apotettix eurycephalus*, and the ensemble of indistinguishable factors responsible for it, are represented by the symbol +/+; (2) a dense black extending over the anterior pronotum and on the femora of the jumping legs is indicated by the letters Bl; (3) a grayish white over the anterior pronotum and lateral lobes, resembling C, of *Paratettix texanus*, although the posterior pronotum and legs are quite yellow, is represented by the letter C; (4) a dark greenish gray over the pronotum, including vertex, eyes and femora, with lighter lateral lobes, ranging almost to white, is designated by the letter H; (5) a dull brick red covering the vertex, eyes, pronotum and posterior femora, is designated by the letter R. A few specimens of +/+ and +/C from Pasadena, California, were later added.

All adult individuals of this species (*Telmatettix aztecus*) have had long wings and pronota, as against *P. texanus* and *A. eurycephalus*, individuals of which frequently have had short wings and pronota (NABOURS, 1914, 1917, 1925, NABOURS and SNYDER, 1928).

The experimental breeding of this species is much more difficult than is that of *P. texanus* and still more arduous than of *A. euryceph-*

alus. A large proportion of the matings are not productive, due usually to the early death of one or both individuals. During the whole period 1922—27, only 116 of the 187 matings made were productive. During the one year of most intensive breeding, May, 1923, to June, 1924, only 53.9 per cent of the matings were productive; the range throughout the months of that year is indicated in table IX.

TABLE XI. MATINGS MADE COMPARED WITH NUMBER PRODUCTIVE-
YEAR 1923—24

Month	Matings Made	Productive Matings
June	4	2
July	2	1
August	5	3
September	18	7
October.	28	13
November	5	1
December	2	1
January	1	1
February	9	3
March	8	3
April.	21	17
May	12	10
Totals	115	62

53.9 per cent productive.

The number of individuals recorded fell far short of the number of young hatched and transferred from the mating jars. During the whole period of five years, only 50.6 per cent or 2850 of the 5634 offspring transferred, were finally recorded. However, during the year June, 1923 to July, 1924, when better attention was given them, 55.19 per cent of those transferred had their color patterns recorded. Table X shows the distribution of the mortality throughout the year.

TABLE X. NUMBER TRANSFERRED COMPARED WITH NUMBER RE-
CORDED — YEAR 1923—24

Month	Numbers Transferred	Numbers Recorded
July	214	110
August	64	30
September	0	0
October.	154	61
November	317	96
December	28	10
January	50	19
February	313	173
March	155	94
April	264	169
May	504	272
June	1,448	904
Totals	3,511	1,938

55.19 per cent recorded.

Segregation of Factors in the Mated Individuals (Males on left, females on right of hyphen). The segregation of the factors of all the individuals heterozygous for the color patterns is shown in Table XI.
Referring to the table: beginning at the left with the item Bl/C, for
example, all males bred gave 166 carrying Bl: 165 carrying C; the
females gave 192 Bl: 174 C; both males and females gave 358 Bl: 339
C, in total of 697 gametes.

TABLE XI. A SUMMARY OF THE SEGREGATION OF FACTORS OF HETEROZYGOUS MALES AND FEMALES OF *Telmatettix aztecus*

Hetero-zygotes	Segregation of factors in males		Segregation of factors in females		Segregation of factors in males and females combined		Total gametes
+/C	+ 153	C 157	+ 213	C 228	+ 366	C 385	751
+/Bl	+ 312	Bl 276	+ 96	Bl 96	+ 408	Bl 372	780
+/H	+ 67	H 53	+ 67	H 60	+ 134	H 113	247
+/R	+ 131	R 121	+ 145	R 167	+ 276	R 288	564
Bl/C	Bl 166	C 165	Bl 192	C 174	Bl 358	C 339	697
Bl/H	Bl 0	H 1	Bl 32	H 24	Bl 32	H 25	57
Bl/R	Bl 0	R 0	Bl 99	R 124	Bl 99	R 124	223
C/H	C 211	H 215	C 96	H 68	C 307	H 283	590
C/R	C 149	R 130	C 58	R 62	C 207	R 182	389
H/R	H 33	R 39	H 24	R 26	H 57	R 65	122
Totals	1222	1147	1022	1029	2244	2176	4420

In addition to the data summarized in the table above, there was one mating of a male +/C to a female +/C. This gave 44 +/C and C/C: 14 +/+ individuals, a close approximation of the 3 : 1 ratio.

Cytology. Cytological preparations for the study of spermatogenesis in *T. aztecus*, prepared with the aid of Doctor MARY T. HARMAN, exhibit six and seven chromosomes in the nuclei of the secondary spermatocyte divisions.

Summary and Conclusions. 1. A series of four factors for dominant color patterns and their normal recessive have been analyzed. It appears that these factors are confined to the one series, or one pair of chromosomes, and that they are alternative, or so closely linked that no crossing over has occurred during the experiments.

2. The chromosome numbers have been found to be six and seven in the secondary spermatocyte division cells.

Tettigidea parvipennis pennata MORSE — *The Color Patterns of T. parvipennis pennata.* (1) A yellowish white striped pronotum was designated as C; (2) two whitish lines extending the full length of the margins of the pronota (bilineata), D; (3) slightly fulvo-aeneous plus blackish striped pronotum, E; (4) narrow white spots, or bands on the posterior femora, F; (5) melanic, i. e., the whole animal a dirty-brown to almost black, M. These five compose a group of apparent alternatives. (6) Light brownish-red pronotum and femora of the jumping legs, H. This one, H, is extremely loosely linked with the five alternatives, or multiple allelomorphs, and it may be of another pair of chromosomes and entirely independent (BELLAMY, 1917, and his Plate).

Directions for Reading Table XII Showing Segregation of Factors for Color Patterns in T. parvipennis pennata. The first column shows the pairs of factors; the next two, under males, give the respective numbers of gametes, recorded from the males. The two middle columns show the gametes of the females, and the last column gives the total number of gametes from the respective pairings. Example, of the 462 gametes produced by individuals carrying D and F, the males gave

TABLE XII. SEGREGATION OF FACTORS FOR COLOR PATTERNS IN
Tettigidea parvipennis pennata (SUMMARIZED FROM BELLAMY'S
TABLE, 1917)

Hetero-zygotes	Males		Females		Males Females		Total gametes
C/D	C	D	C	D	C	D	
	10	9	8	6	18	15	33
C/E	C	E	C	E	C	E	
	0	0	3	4	3	4	7
C/F	C	F	C	F	C	F	
	78	79	0	0	78	79	157
C/M	C	M	C	M	C	M	
	96	82	0	0	96	82	178
D/E	D	E	D	E	D	E	
	4	5	165	156	169	161	330
D/F	D	F	D	F	D	F	
	180	198	46	38	226	236	462
D/M	D	M	D	M	D	M	
	25	17	59	62	84	79	163
E/F	E	F	E	F	E	F	
	32	47	85	101	117	148	265
F/M	F	M	F	M	F	M	
	22	28	43	48	65	76	141
Totals	447	465	409	415	856	880	1736

180 D: 198 F; the females 46 D : 38 F; both males and females gave
226 D : to 236 F.

*Explanation of the Use of Table XIII Showing Relations of the Factor
H with the Alternatives C, D, E and F, in Males and Females of T. par.
pennata.* The pairings are shown at the left. The first column of figures shows the numbers, and the second column the crossovers, or
segregations; the third column gives the percentage of crossing over,
or segregation, in the males. The three columns, to the right, give the
same data for the females. There were no pairings of C and H in the

females. Example, in the males, there were 355 gametes from EH pairings of which 171, 48.17 per cent ,were crossovers or independent. In the females there were 59 gametes from EH pairings with 35 or 59.32 percent of independent segregation, or crossing over. The totals show a crossing over, or independent segregation of 45.81 per cent in the males and 47.95 per cent in the females.

TABLE XIII. SHOWING RELATIONS OF THE FACTOR H WITH THE ALTERNATIVES C, D, E AND F IN THE MALES AND FEMALES OF *Tettigidea parvipennis pennata*

	Males			Females		
Factors	Numbers	Cross-overs [1]	Percent of cross-ing over	Numbers	Cross-overs [1]	Percent of cross-ing over
CH	14	6	42.85			
DH	312	139	44.55	39	12	30.76
EH	355	171	48.17	59	35	59.32
FH	251	111	44.22	98	47	47.95
Total	932	427	45.81	196	94	47.95

Discussion and Summary of the Results of Breeding T. parvipennis pennata. The factors for the color patterns designated as C, D, E, F and M are certainly extremely closely linked, if they are not alternatives, or a series of multiple allelomorphs. The gene for the color pattern represented by H is probably located in a different pair of chromosomes from the five alternatives, as there is close to 50 per cent segregation in both the males and females (BELLAMY, 1917; HALDANE, 1920).

PARTHENOGENESIS IN THE TETTIGIDAE

The species *Paratettix texanus*, *Apotettix eurycephalus*, *Telmatettix aztecus* and *Tettigidea parvipennis pennata*, have produced offspring

---

[1] Probably independent segregation.

parthenogenetically in the breeding experiments (NABOURS, 1919, 1925; NABOURS and FOSTER, 1925, 1929; NABOURS and SNYDER, 1924, 1928). Bisexual reproduction appears to be the normal method, but if the males are impotent, die before copulation, or if the females are not mated at all, a limited number of offspring, preponderantly females, are hatched from the unfertilized eggs. Rarely, if ever, do ɟemales give parthenogenetic offspring when they are mated to potent males. It appears that males are more frequently impotent than females. Examinations of the reproductive organs have revealed a number of cases of absence or deficiency in numbers of spermatozoa.

Parthenogenesis in P. texanus. Seventy-five females which were not exposed to males gave 625 offspring of which 393 females and one male became large enough to be recorded and there were seven the sex of which was not noted. In addition, thirty-three females, which had been exposed to males of contrasting dominant color patterns, produced offspring of which 187 females and one male were recorded. The offspring from these exposed females (1) were preponderantly females, (2) exhibited no traces of the dominant color patterns of the males with which the females had been placed, and, whenever tested by further breeding, (3) showed themselves to have been homozygous for those factors for which they should have been heterozygous had the eggs been fertilized; therefore, they were thought to have been from unfertilized eggs (NABOURS and FOSTER, 1925, 1929).

Besides these 580 recorded females and two males which were produced parthenogenetically, fifteen mated females, in addition to their bisexual progenies, each gave from one to seven offspring, all females, which did not show any influence of the dominant characteristics of the males. They thus satisfied two of the criteria for parthenogenesis (loc. cit.).

Parthenogenesis in A. eurycephalus. There had been 5313 females and thirteen males hatched from unfertilized eggs in *A. eurycephalus* when the data were last assembled. Tables and descriptions showing the details of these experiments were published by NABOURS (1925). There were many females which were never exposed to males, some of them producing parthenogenetically for seven consecutive generations, and a few with which males were placed, but all the criteria pointed to the fact of the parthenogenetic origin of all their progenies (loc. cit.). As in *P. texanus*, a few females gave offspring both bisexually and parthenogenetically, and there were some which gave offspring from unfertilized eggs, and then, after mating with males, bisexual progenies.

Parthenogenesis in Telmatettix aztecus. Among fifteen groups of females which had not been exposed to males after becoming adult, twelve gave from one to 112 offspring. From two females, +/C, in one cage, 112 offspring hatched, and 102 were recorded, of which 49 were +/+ and 53 C/C. Twelve females heterozygous for two factors, respectively, gave 68 of the one and 70 of the other alternative (Items 8—12, Table IV, NABOURS and SNYDER, 1928).

The individuals which gave parthenogenetic offspring were from progenitors collected exclusively from a small area along Shoal Creek, Austin, Texas, and had not been hybridized in the laboratory with stock from any other place.

Parthenogenesis in Tettigidea parvipennis pennata. There have been a few *T. par. pennata* females produced by parthenogenesis in the Kansas Agricultural Experiment Station laboratory. One unmated individual of this species, of the genetic composition F/M, for color patterns (BELLAMY, 1917), gave two F/F and three M/M females (NABOURS and FOSTER, 1929).

Summary and Conclusions Regarding Parthenogenesis in the Tettigidae. The members of the species *P. texanus*, *A. eurycephalus*, *T. aztecus* and *T. parvipennis pennata* have been found to be bisexual, the fertilized eggs producing males and females in equal numbers, and parthenogenetic, the unfertilized eggs, with rare exceptions, hatching females. The females are generally much more prolific when mated than when unmated. A mated female may have part of her ova fertilized, and also produce from unfertilized ova, by parthenogenesis, an additional number of offspring which are nearly always females.

The segregation of factors occurs in all the individuals producing parthenogenetically apparently to the same extent as in those reproducing bisexually. Crossing over of factors occurs in the females, where there is a linear arrangement of the genes, as in *P. texanus* and *A. eurycephalus*, in parthenogenetic breeding apparently to the same extent as in bisexual reproduction.

The inheritance results in parthenogenesis indicate that segregation and crossing over occur before the inception of the parthenogenetic processes, and that up to this stage in gametogenesis there is no essential departure from the usual procedure in bisexual reproduction.

As previously noted, p. 53, when the full number of fourteen chromosomes of the soma and oögonial cells of parthenogenetically produced individuals of *P. texanus* and *A. eurycephalus* are manifest, the members of the homologous pairs, respectively, lie together in early cell divisions, and not far apart, each from the other, in later cell generations, in such positions as to suggest that the second polar body division had been inhibited (ROBERTSON, 1929, MSS.).

The second polocyte division in the *Tettigidae* (loc. cit.) as in other forms (see WILSON, 1925), is probably normally consequent upon the entrance of the sperm, another case of a later stage of maturation being overlapped by an earlier stage of fertilization. In the absence of the fertilizing sperm and the resultant complete, or partial inhibition of the last polocyte division, the diploidal condition is retained or restored, and if the specific gene, or complementary genes responsible for the parthenogenetic processes are present (NABOURS and FOSTER, 1929), a chemical situation arises which conditions the initiation of development. Since it appears that any egg of these species is capable of being fertilized (those without the specific genes require it), such educement of development as described is, in effect, *artificial* parthenogenesis.

The parthenogenesis exhibited by these grouse locusts can hardly be classed as *facultative* and certainly not as *obligatory* (see definitions, WILSON, 1925, pp. 228, 229). It might perhaps be best entitled *tychoparthenogenesis* and preponderantly *gynogenetic*.

BATESON's (1922) challenge that a new stock must be evolved, or observed to evolve, which will not cross back with the parent stock, or if it does, will produce only sterile progeny, before evolution can be spoken of as a *fait accompli*, has been and appears still to remain the *impedimentum supremum* in experimental evolution. This impediment was duly noted by DARWIN and others. If a germinal variation or mutation of sufficient range to overcome this barrier should occur, or be educed in only the one individual, of either sex, it would be useless. The chance that the same variation would occur in the two sexes coincidentally is so remote as to render the idea of it untenable. Not as a general rule, but to cover a special way, if a sufficient mutation should occur in a female of tychoparthenogenetically reproducing organisms, might not this difficulty be obviated? In one line, among several, a female of *P. texanus* exposed to a male gave progeny parthenogenetically. Seven of these offspring were mated; five gave bisexual progenies, but the other two again reproduced parthenogenetically in spite of the males to which they were exposed. However, in this case the progeny of the second parthenogenetic generation, exposed to males, mated and gave bisexual offspring (NABOURS and FOSTER, 1929). The two successive parthenogenetic generations, while the females were exposed to males, had very likely been a matter of impotency of the males rather than a mutation in a female. Nevertheless, there may be possibilities in this direction.

PEACOCK and HARRISON (1925) found that hybrid moths from the crossing of *Tephrosia bistortata* males with *T. crepuscularia* females reproduced parthenogenetically. The unmated females of neither species alone would reproduce. They concluded mainly on the basis of this result *that parthenogenesis was consequent upon hybridity*. In a later paper (1926) these authors adduced from my tables (NABOURS, 1919, 1925) that all the parthenogenetic females of the grouse locust *A. eurycephalus* had been hybrids from the crossing of one variety of this species from the region of Houston, Texas, with another from Tampico, Mexico. They welcomed this as evidence constituting strong confirmation of their hypothesis. This proposition of PEACOCK and HARRISON has probably received further support from the results of the parthenogenetic breeding of *P. texanus* (p. 89, and NABOURS and FOSTER, 1929), if it be provided, as PEACOCK and HARRISON did not, *that the process of hybridism may bring together specific complementary, or climaxing genes which are responsible for, or cause the development of the unfertilized eggs.*

INHERITANCE IN ORTHOPTERA OTHER THAN THE
TETTIGIDAE

The difficulties of breeding the *Locustidae, Acrididae* and the *Gryllidae* have made the members of these families largely unavailable for inheritance studies. The few species of the *Acrididae* and *Locustidae* with which breeding experiments have been attempted, were extremely susceptible to fungus and other diseases and required from one to three years for each generation. The variety of color of the contrasting characteristics is not nearly so great in any of these as in several species of the sub-family *Tettigidae*.

HANCOCK's *Studies of Inheritance in Green and Pink Katy-dids, Amblycorypha oblongifolia* DE GEER

A green male was mated to a pink female Katy-did, both of the species *Amblycorypha oblongifolia*. There were thirteen F_1 progeny, eight pink and two green hatched two years, and one pink and two green hatched three years after the eggs had been oviposited. The F_1 and F_2 greens, as well as green parents from the field, when inbred, invariably gave green progenies. The F_1 pink individuals, from green mated to pink, when inbred, produced, requiring two and three years for hatching, 38 green: 90 pink offspring. It thus appears that the green and pink colors of *A. oblongifolia* compose a pair of Mendelian characters, with the pink color dominant (HANCOCK, 1916; NABOURS, 1928).

CAROTHERS's *Identification of the Homologues of Given Pairs of Chromosomes from Parents to Offspring in Circotettix verruculatus*

CAROTHERS (1921) was able to make actual identification of the homologues of given pairs of chromosomes, the members of the heteromorphic, telomorphic and atelomorphic pairs, from fathers to sons in genetical experiments with *Circotettix verruculatus*. These well-marked chromosomes segregated at random in such a way as to parallel the behavior of unit characters in MENDEL's first and second laws of inheritance.

The Phasmidae

Several persons have bred the walking sticks, mostly the species *Bacillus rossii* and *Carausius* (*Dixippus*) *morosus*, which reproduce both bisexually and parthenogenetically. MacBride and Jackson (1915) found that the females of *C. morosus* produced males and females in the ratio of about one to five hundred in parthenogenesis. They reported that either brown or green individuals gave a ratio of about one brown to 25 green progeny even in the third and fourth parthenogenetic generation. They concluded that the color characteristics were not influenced by the environment. Fryer (1914) worked with the bisexual walkingstick, *Clitumnus cuniculus*, from Ceylon and was led to believe the color differences were inherited.

On the other hand, Dobkiewicz (1912) and Mangelsdorf (1926) interpreted the variations in colors of *Dixippus* (*Carausius*) *morosus* as due to the influence of the environment. Pantel and de Sinety (1918) came to the conclusion that other species of walking-sticks were dependent upon the surroundings for color variations.

Wankel (1871) and Graber (1873) observed polyandry and polygamy in the *Tettigidae* and other grasshoppers. Bolivar (1897), Dominique (1897, 1901), de Sinéty (1901), Godelmann (1901), Hanitsch (1902), Daiber (1904), Polak (1904), Stockard (1909), Nachtsheim (1923) and Cappe de Baillon (1926) are among many who have experimented with, or observed parthenogenesis and inheritance in the *Phasmidae*. The changes of colors from the time of hatching till the adult stage, the great mortality among the immature stages, time required for each generation, and other complications have rendered these insects very difficult material with which to carry on experimentation.

The Mantidae

Mr. E. H. Ingersoll (1923—26, unpublished), in the Kans. Agricultural Experiment Station Laboratory, bred a considerable number of individuals of *Stagmomantis carolina*. These exhibited a variety of color patterns. He was able to determine that the nymphal colors were not influenced by the environment. Green appeared as a recessive to several other colors. Avellaneous body color was dominant to a dark cinnamon pink. These insects proved to be extremely difficult to breed, and their color characteristics, including the nymphal changes, were extraordinarily complex.

PRZIBRAM (1906, 1909), RAU and RAU (1913), ADAIR (1924,) and RABAUD (1926) have studied color variations, parthenogenesis and, to some extent, heredity among the species of the *Mantidae*.

LITERATURE

ADAIR, E. W. 1924. On Parthenogenesis in *Miomantis savignyi* SAUSS. Bulletin Royal Soc. Entomology of Egypt, Cairo, 1924, pp. 104—148.

BATESON, W. 1922. Evolutionary Faith and Modern Doubt. Science. Vol. 55, pp. 55—61.

BELLAMY, A. W. 1917. Multiple Allelomorphism and Inheritance of Color Patterns in *Tettigidea*. Journ. Genetics, 7, pp. 55—70.

BOLIVAR, IGNACIO. 1887. Essai sur les *Acridiens* de la tribu des *Tettigidae*. Ann. Soc. Ent. Belg. 31, pp. 175—313.

BOLIVAR, IGNACIO. 1897. La parthenogenesis en los ortopteros. Act. Soc. Esp. Hist. Nat., 1897.

CAPPE DE BAILLON, P. 1926. Variation et Parthénogénèse. Note sur la biologie de quelques Phasmides. Bull. Biol. France et Belg., Paris. 60, pp. 474—482.

CAROTHERS, E. ELEANOR. 1913. The Mendelian Ratio in Relation to certain Orthopteran Chromosomes. Jour. Morph., 24, pp. 487—511.

CAROTHERS, E. ELEANOR. 1917. The Segregation and Recombination of homologous Chromosomes as found in two Genera of *Acrididae*. Jour. Morph., 28, pp. 445—521.

CAROTHERS, E. ELEANOR. 1921. Genetical Behavior of heteromorphic homologous Chromosomes of *Circotettix*. (Orthoptera) Jour. Morph., 35, pp. 457—483.

CASTLE, W. E. 1914. NABOURS' Grasshoppers, multiple Allelomorphism, Linkage and misleading Terminologies in Genetics. Am. Nat., Vol. 48, pp. 383—384.

DAIBER, MARIE. 1904. Beitrage zur Kenntnis der Ovarien von *Bacillus rossi* nebst einigen biologischen Bemerkungen. Jenaische Zeitschrift für Naturwissenschaft. 39, pp. 177—202.

DEXTER, J. S. 1914. NABOURS's Breeding Experiments with Grasshoppers. Amer. Nat., 48, pp. 317—320.

DEXTER, J. S. 1918. Inheritance in *Orthoptera*. Amer. Nat., 52, pp. 61—64.

DOBKIEWICZ, L. VON. 1912. Einfluss der äusseren Umgebung auf die Farbung der indischen Stabheuschrecken *Dixippus morosus*. Biol. Zentralbl., 32, pp. 661—663.

DOMINIQUE, J. 1897. Parthénogénèse et Thelytokie chez les *Phasmides*. Bull. Soc. Ouest France, IX, pp. 127—136.

DOMINIQUE, J. 1901. Encore quelques mots sur l'élévage des *Bacilles*. Bull. Ouest France, XI, pp. 229—234.

FRYER, J. C. F. 1914. Preliminary Note on some Experiments with a Polymorphic Phasmid. Jour. Genetics, 3, pp. 107—111.

GODELMANN, R. 1901. Beiträge zur Kenntnis von *Bacillus rossii* FABR. Arch. für Entwickmechan. der Organismen, 12, pp. 265—301.

GRABER, V. 1873. Ueber Polygamie und anderweitige Geschlechtsverhältnisse bei Orthopteren. Verhandlungen der Zoologisch-Botanischen Gesellschaft, in Wien., 21, pp. 1091—1096.

HAAN, WILLIAM DE. 1842. Bijdr. Kenn. *Orthoptera*. (First synoptical table of the species of the *Tettigidae*), 7, pp. 166.

HALDANE, J. B. S. 1920. Note on a Case of Linkage in *Paratettix*. Jour. Genetics, 10, pp. 47—51.

HANCOCK, J. L. 1898. The Food Habits of the *Tettigidae*. Ent. Rec. and Jour. of Variation, 10, pp. 6, 7.

HANCOCK, J. L. 1899. Some Tettigian Studies. Ent. News, 10, pp. 275—282.

HANCOCK, J. L. 1900. Notes on Species of the Tettigian Group of *Orthoptera*. Can. Ent., 32, pp. 25—27.

HANCOCK, J. L. 1900. A new Tettigian Genus and Species from South America. Psyche, 6, pp. 42, 43.

HANCOCK, J. L. 1900. Synopsis of Subfamilies and Genera of North American *Tettigidae*. Psyche, 9, pp. 6, 7.

HANCOCK, J. L. 1902. The *Tettigidae* of North America. Chicago, pp. 1—188. (Extensive bibliographies).

HANCOCK, J. L. 1904. The *Tettigidae* of Ceylon. Spolia Zeylanica, Vol. II, part 7, pp. 97—157.

HANCOCK, J. L. 1906. On the Orthopteran Genus *Ageneotettix*, with a Description of a new Species from Illinois. Ent. News, 17, pp. 253—256.

HANCOCK, J. L. 1906. Descriptions of new Genera and Species of the Orthopterous Tribe *Tettigidae*. Ent. News, March, 1906, pp. 86—91.

HANCOCK, J. L. 1907. Studies of the *Tetriginae* (*Orthoptera*) in the Oxford Museum. Trans. Ent. Soc., London. Sept. 26, 1907. pp. 213—244.

HANCOCK, J. L. 1908. A new Ceylonese Tettigid (*Orthoptera*) of the Genus *Eurymorphopus*. Spolia Zeylanica, 5, part 19, pp. 113—14.

HANCOCK, J. L. 1909. Further Studies of the *Tettiginae* (*Orthoptera*) in the Oxford Museum. Trans. Ent. Soc., London. Jan. 20, 1909. pp. 387—426.

HANCOCK, J. L. 1910. Notes on Ceylonese *Tettiginae* (*Orthoptera*), with Descriptions of some new Species. Spolia Zeylanica, 6, part 24, pp. 140—149.

HANCOCK, J. L. 1912. *Tetriginae* (*Acridiinae*) in the Agricultural Research Institute, Pusa, Bihar, with Descriptions of new Species. Mem. Dept. Agric., India, Ent. Series, 4, pp. 131—160.

HANCOCK, J. L. 1916. Pink Katydids and the Inheritance of Pink Coloration, Ent. News, 27, pp. 70—82.

HANITSCH, R. 1902. On the parthenogenetic Breeding of *Eurycnema herculeana*. Journal Straits Asiatic Soc., No. 38, p. 35.

HARMAN, MARY T. 1915. Spermatogenesis in *Paratettix*. Biol. Bull. 29, pp. 262—277.

HARMAN, MARY T. 1920. Chromosome Studies in *Tettigidae*. II. Chromosomes of *Paratettix* BB and CC and their Hybrid BC. Biol. Bull., 38, pp. 213—231.

HARMAN, MARY T. 1925. The reproductive System of *Apotettix eurycephalus*. Jour. Morph. and Phys., 41, pp. 217—237.

JANSSENS, F. A. 1909. La théorie de la chiasmatypie. Nouvelle interprétation des cinéses de maturation. La Cellule, tom. 25.

LATRIELLE, P. A. 1804. *Orthoptera*. Vol. XII (First description of *Tettix*).

LEWIS, MARGARET REED, ans W. R. B. ROBERTSON. 1916. The Mitochondria and other Structures observed by Tissue Culture Method in the male Germ Cells of *Chorthippus curtipennis*. Biol. Bull. 30, pp. 99—124.

LINNÉ, CARL VON. 1767. Syst. Nat. (Twelfth Edition, 1767).

MACBRIDE, E. W., and Miss A. JACKSON. 1915. Inheritance of Color in the Stick Insect *Carausius morosus*. London Proceed. Royal Society (Biology), 89, pp. 109—118.

MANGELSDORF, A. J. 1926. Color and Sex in the Indian Walking-stick, *Dixippus morosus*. Psyche, 33, pp. 151—155.

McCLUNG, C. E., 1902. The accessory Chromosome Sex Determinant. Biol. Bul. 3.

McCLUNG, C. E. 1914. A comparative study of the chromosomes in Orthopteran spermatogenesis. Jour. Morph., 25, pp. 651—749.

MORGAN, T. H., A. H. STURTEVANT, H. J. MULLER, and C. B. BRIDGES. 1915. The Mechanism of Mendelian Heredity. Holt and Co., N. Y.

MORGAN, T. H., 1919. The physical Basis of Heredity. 305 pp. J. B. Lippincott Co., Phila.

MORGAN, T. H., A. H. STURTEVANT and C. B. BRIDGES, 1920. The Evidence for the linear Order of the Genes. Proc. Nat. Acad. Sci. 6 : No. 4, pp. 162—164.

MORGAN, T. H., C. B. BRIDGES and A. H. STURTEVANT, 1925. The Genetics of *Drosophila*. Bibliographia Genetica II, pp. 1—262.

MULLER, H. J., 1920. Are the Factors of Heredity arranged in a Line? Am. Nat. 54 : 97—121.

MORSE, A. P. 1900—01. *Tettiginae*. Biol. Cent. Am., Vol. II.

NABOURS, ROBERT K. 1914. Studies of Inheritance in *Orthoptera*. I. *Paratettix texanus*. Jour. Genetics, 3, pp. 141—170.

NABOURS, ROBERT K. 1917. Studies of Inheritance in *Orthoptera*. II and III. *Paratettix texanus* and a Mutant. Jour. Genetics, 7, pp. 1—54.

NABOURS, ROBERT K. 1919. Parthenogenesis and Crossing Over in the Grouse Locust *Apotettix*. Am. Nat., 53, pp. 131—142.

NABOURS, ROBERT K. 1923. A new dominant Color Pattern and Combinations that breed true in the Grouse Locusts. Genetica, 5, pp. 477—480.

NABOURS, ROBERT K. 1925. Studies of Inheritance and Evolution in *Orthoptera*. V. The grouse locust *Apotettix eurycephalus* HANCOCK. Kans. Agri. Exp. Sta. Tech. Bull. 17, pp. 1—231.

NABOURS, ROBERT K. 1927. Polyandry in the Grouse Locust, *Paratettix texanus* HANCOCK, with notes on the Inheritance of acquired Characters and Telegony. Am. Nat., 61, pp. 531—538.

NABOURS, ROBERT K. 1928. HANCOCK's Studies of Inheritance in green and pink Katy-dids, *Amblycorypha oblongifolia* DE GEER. Ent. News, vol. 40, pp. 14—16.

NABOURS, ROBERT K. and BERTHA SNYDER. 1924. Inheritance of Color Patterns in the Grouse Locust *Telmatettix aztecus* SAUSSURE. (Abstract) Anat. Rec., 29, p. 152.

NABOURS, ROBERT K. and BERTHA SNYDER. 1928. Parthenogenesis and inheritance of Color Patterns in the Grouse Locust, *Telmatettix aztecus* SAUSSURE. Genetics, 13, pp. 126—132.

NABOURS, ROBERT K. and MARTHA FOSTER. 1929. Parthenogenesis and the Inheritance of color Patterns in the Grouse Locust *Paratettix texanus*. Biol. Bull., Vol. LVI.

NACHTSHEIM, H. 1923. Parthenogenese, Gynandromorphismus und Geschlechtsbestimmung bei Phasmiden. Zeitschr. f. Indukt. Abst. und Vererbgsl., 30, pp. 287—289.

PANTEL, J. and R. DE SINÉTY. 1918. Réaction chromatique et non chromatique de quelques phasmides. Bull. Biol. de France et Belg. 52, pp. 177—283.

PEACOCK, A. D. and J. W. HESLOP HARRISON. 1925. On Parthenogenesis originating in Lepidopterous Crosses. Trans. Nat. Hist. Soc., Northumberland, Durham and New Castle-upon-Tyne, N. S., 6, Part II.

PEACOCK, A. D. and J. W. HESLOP HARRISON. 1926. Parthenogenesis and Segregation. Nature, 117, pp. 378, 379.

POLAK. 1904. Entomologische Berichten .Uitgegeven door de Nederlandsche entomologische Vereeniging, 1, p. 148.

PRZIBRAM, HANS. 1906. Aufzucht Farbwechsel und Regeneration einer ägyptischen Gottesanbeterin (*Sphodromantis bioculata* BURM.). Archiv für Entwicklungsmechanik der Organismen, 22, pp. 149—206.

PRZIBRAM, HANS. 1909. Aufzucht Farbwechsel und Regeneration der Gottesanbeterinnen (*Mantidae*). III. Temperatur- und Vererbungsversuche Arch. für Entwicklungsmechanik der organismen, 26, pp. 561—628.

RABAUD, ÉTIENNE. 1914. Telegony. Jour. Heredity. Vol. 5, pp. 389—399.

RABAUD, ÉTIENNE. 1926. Variation chromatique chez *Mantis religiosa*. Soc. de Biol. Comptes Rendus, 94, pp. 36, 37.

RAU, PHIL and NELLIE RAU. 1913. The Biology of *Stagmomantis carolina*. The Acad. of Sci. of St. Louis Trans. 22, pp. 1—58.

RAYBURN, MYRTLE F. 1917. Chromosomes of *Nomotettix*. Kans. Univ. Sci. Bul., 10, pp. 267—270.

REHN, JAS. A. 1901. Mexican *Orthoptera*. Trans. Am. Ent. Soc., 27, p. 229.

ROBERTSON, W. R. B. 1908. The Chromosome Complex of *Syrbula admirabilis*. Kans. Univ. Sci. Bul., 4, 275—305.

ROBERTSON, W. R. B. 1915. Chromosome Studies. III. Inequalities and Deficiencies in homologous Chromosomes; their Bearing upon Synapsis and the Loss of Unit Characters. Jour. Morph., 26, pp. 109—141.

ROBERTSON, W. R. B. 1916. Chromosome Studies. I. Taxonomic Relationships shown in the Chromosomes of *Tettigidae* and *Acrididae*: V-shaped Chromosomes and their Significance in *Acrididae*, *Locustidae* and *Gryllidae*: Chromosomes and Variations. Jour. Morph., 27, pp. 179—280.

ROBERTSON, W. R. B. 1917. Chromosome Studies. IV. A deficient supermumerary accessory Chromosome in a Male of *Tettigidea parvipennis*. Kansas Univ. Sci. Bull. 10, pp. 275—283.

ROBERTSON, W. R. B. 1925. The chromosomes of Dr. NABOURS' parthenogenetically produced *Tettigidae*. Anatomical Record, 31, pp. 307, 308.

ROBERTSON, W. R. B., 1929. Chromosome Studies V: The Retention of Diploidy in Partheno-produced *Tettigidae*, including two rare Males, and the Tendency of Chromosomes to persist in their original gametic Positions in the Cells of these and their Biparentally produced Relatives. (*Apotettix eurycephalus* and *Paratettix texanus*) (In Manuscript).

SAUSSURE, HENRY DE. 1859—61. *Orthoptera* Nova Americana. Revue et Mag. de Zool. Series I—III, 1859—61.

SCUDDER, S. H. 1900. Catalog of the described *Orthoptera* of the U. S. A. and Canada. Proceed. Davenport Acad. Sci., Vol. VIII.

SINÉTY, R. DE. 1901. Recherches sur la biologie et l'anatomie des Phasmes. La Cellule, 19, pp. 118—278.

STOCKARD, C. R. 1909. Inheritance in the Walking Stick, *Aplopus mayeri*. Biol. Bull. 16, pp. 239—245.

SUTTON, W. S. 1902. On the Morphology of the Chromosome Group in *Brachystola magna*. Biol. Bull., 4, No. 1.

VOINOV, D. N. 1914a. Sur un nouveau mécanisme déterminant le dimorphisme des éléments sexuels; chromosome et polarité variable. C. R., Soc. Biol., Paris, T. 76.

VOINOV, D. N. 1914b. Recherches sur la spermatogénèse du *Gryllotalpa vulgari*. Arch. de Zool. Expt. et gen. T. 54.

WANKEL, CARL. 1871. Orthopterologische Studien. Zeitschrift für die gesammten Naturwissenschaften. 1871, pp. 1—28.

WENRICH, D. H. 1914. Synapsis and the Individuality of the Chromosomes. Science, N. S., 41.

WENRICH, D. H. 1916. The Spermatogenesis of *Phrynotettix magnus* with special Reference to Synapsis and the Individuality of the Chromosomes. Bull. Mus. Comp. Zool., Harvard College, Vol. 60.

WILSON, E. B. 1925. The Cell in Development and Heredity. Macmillan Co., New York. 1232 pp.

INDEX

Abdominal appendages of male 30.
Accessory chromosome 54.
Accessory glands 44.
Acrydium granulatus 30, 32, 48 *hancocki* 30, *obscurus*, nymph of 30; *obscurus* 32, 48; *ornatus triangularis* 49.
ADAIR 95, 96.
Adults, hibernation of 34.
Alternation of hibernating generations 34.
Amblycorypha oblongifolia, inheritance in 93.
Anatomy, external 29.
Apical process 31, 32.
Apotettix eurycephalus 32, 48, 49, 50, 55, 76; better breeders 36; experimental breeding 78, 81; parthenogenesis in 89, 90; reproductive systems 43.
Appendages, male sexual 31.
Artificial parthenogenesis 91.

Bacillus rossii 94.
Basal cyst 44.
BATESON 91, 96.
BATESON's challenge 91, 92.
Batrachoseps 53.
BELLAMY 35, 55, 86, 96.
Biology 29.
BOLIVAR 94, 96.
Breeding difficulties 35, habits 33; in the greenhouse 35; optimum season 35, 36; favorable months for 38.
BRIDGES 82, 98.
Bulla of LINNÉ 29.
Burrow, preparation for ovipositing 30.

Caeca 32.
Cages, kinds of 36.
CAPPE DE BAILLON 94, 96.
Carausius (Dixippus) morosus 94.
CAROTHERS 55, 93, 96.
Cerci 44.
Characteristics 29, 55—58, 77, 78, 82, 90.
Chiasmatype theory 53.
Choriphylum foliatum 30, 51.
Chorthippus 53.
Chromosomes and Mendelian Laws 55.

Chromosomes in parthenogenesis 46, 47; definite organization 54; general consideration 54; map making 78, 79; relations in parthenogenesis 53; of *Acrydium* 52; *Apotettix eurycephalus* 52; of bisexually produced females 52; of brain cell 46, 47; of *Choriphylum* 51; of *Gryllidae* 54; of *Nomotettix* 51; of oögonium cell 46, 47; of ovarian follicle wall cell 46, 47; of *Paratettix* 52; of partheno-produced females 53; of partheno-produced males 53; of spermatogonium 46, 47; of *Telmatettix aztecus* 86; of *Tettigidea* 52; of the *Acrididae* 54; of the *Blattidae* 54; of the *Forficulidae* 54; of the *Locustidae* 54; of the *Phasmidae* 54; of the *Tettigidae* 51; perpetuation of 54.
Circotettix verruculatus, inheritance in 93.
Clitumnus cuniculus 94.
Cold resistance 34.
Collaterial glands, lack of 44.
Collections from Austin, Texas 35; from Houston, Texas 35; from Many, Louisiana 35; from Pasadena, California 35; from San Antonio, Texas 35; from Sugarland, Texas 35; from Tampico, Mexico 35.
Color characteristics and environment 55, 94.
Color patterns 32; combining 32; development of 49; number of elementary 32; of *A. eurycephalus*, description of 77, 78; of *P. texanus*, description of 57, 58; unchanged by environment 55; variations 55; preservation of 43.
Colors and environment 94.
Convoluted tubes 44.
Coupling and repulsion 53.
Crop 32.
Crossing over 53; of factors in parthenogenesis 91; in *P. texanus* 71—75; in *Apotettix eurycephalus* 78—79; in males 70—75.
Cysts of follicles 43.